Impulsive Synchronization of Complex Dynamical Networks

Ze Tang · Dong Ding · Yan Wang · Zhicheng Ji ·
Ju H. Park

Impulsive Synchronization of Complex Dynamical Networks

Modeling, Control and Simulations

Springer

Ze Tang
School of Internet of Things Engineering
Jiangnan University
Wuxi, Jiangsu, China

Dong Ding
School of Internet of Things Engineering
Jiangnan University
Wuxi, Jiangsu, China

Yan Wang
School of Internet of Things Engineering
Jiangnan University
Wuxi, Jiangsu, China

Zhicheng Ji
School of Internet of Things Engineering
Jiangnan University
Wuxi, Jiangsu, China

Ju H. Park
Yeungnam University
Kyongsan, Korea (Republic of)

ISBN 978-981-16-5382-7 ISBN 978-981-16-5383-4 (eBook)
https://doi.org/10.1007/978-981-16-5383-4

This Springer imprint is published by the registered company Springer Nature Singapore Pte Ltd.
The registered company address is: 152 Beach Road, #21-01/04 Gateway East, Singapore 189721, Singapore

This book is dedicated to our families. With tolerance, patience, and wonderful frame of mind, they have encouraged and supported us for many years.

Preface

In the concept of network science, a complex dynamical network is a connected graph with nontrivial topological characteristics, like high cluster effects, the different degree distributions of each node, homogeneous and heterogeneous among different nodes, different types of hierarchical structures and different community structures. These characteristics do not always appear in simple graphs like random graphs and lattices, but often appear in some graphs when modelling the real systems. As one of the most important collective behaviors, the investigation on synchronization of the complex networks has been lasting for several decades until now. In fact, the synchronization of complex networks has been applied in many fields, such as medical treatment, parallel image grabbing and processing, scientific index networks, smart grid and so on. Specially, due to the importance not only in the theoretical analysis but also in practical applications, the study on impulsive synchronization of complex dynamical networks has attracted engineers and scientists from various disciplines, such as electrical engineering, mechanical engineering, mathematics, network science and system engineering.

Pursuing a holistic approach, this book introduces emergent problems, models and issues in impulsive synchronization of complex networks. It establishes a fundamental framework for this topic, while emphasizing the importance of network synchronization and the significant influence of impulsive control in the design and optimization of complex networks. This book is mainly focused on the global impulsive synchronization of complex dynamical networks with different types of couplings, such as general state coupling, nonlinear state coupling, time-varying delay coupling, derivative state coupling, proportional delay coupling and distributed delay coupling. Correspondingly, different types of control methods, especially the impulsive control and pinning control will be introduced in order to derive sufficient conditions for the global synchronization. The primary audience for the book would be the scholars and graduate students whose research topics including the network science, control theory, applied mathematics, system science and so on. The prerequisite knowledge of this book could be the differential dynamic systems, the complex systems theory, the advanced and modern control theory and some basic knowledge of mathematics such as linear algebra and matrix theory.

The main outline of this book is to introduce some recent research works on analysis, control and applications of impulsive synchronization on complex networks. The book is organized as follows:

Chapter 1: The background of synchronization on complex dynamical networks is introduced, as well as the organization of this book, and some important definitions and useful lemmas are also provided in this chapter.

Chapter 2: This chapter aims to study the cluster synchronization for a kind of dynamical complex networks containing nonidentical nonlinear Lur'e systems and proportional delay, which has the features of unbounded and time-varying. Considering different functions that the impulsive effects play, some sufficient criteria for the cluster synchronization of the nonidentically coupled Lur'e dynamical networks are derived by applying the extended parameters variation formula, the impulsive comparison principle and the concept of average impulsive interval.

Chapter 3: This chapter discusses the global exponential synchronization for a class of delay derivative coupled neural networks with multiple time-varying delays and stochastic disturbance. To broaden the fields of synchronization applications in network science, consider cluster-tree topology structure of the coupled neural networks, a novel impulsive pinning control strategy is proposed, which skillfully considered the neural networks in current cluster that directly linked to the neural networks in other clusters.

Chapter 4: This chapter studies the adaptive control and the exponential synchronization problem of a kind of derivative coupled complex dynamical networks with proportional delay. Sufficient criteria of the exponential synchronization on complex networks are given based on impulsive control and adaptive pinning control strategies jointly applying the proportional delayed impulsive comparison principle, the extended parameters variation formula and the definition of average impulsive interval.

Chapter 5: This chapter concentrates on the exponential synchronization of nonlinear coupled Lur'e networks with proportional delay. For the sake of the heterogeneities existed in different Lur'e systems, the investigation on the global quasi-synchronization instead of complete synchronization for the coupled Lur'e networks is given. Comparing to pervious general time delay, as a type of unbounded time-varying delays, the proportional delay largely increases the challenge on network synchronization.

Chapter 6: This chapter is mainly focused on the global and exponential synchronization problem for a class of complex dynamical networks with nonidentically coupling and time-varying delays. Since the mismatched parameters in different systems, the quasi-synchronization issue instead of complete synchronization is investigated based on the application of impulsive control schemes. In view of the concept of average impulsive interval, the extended comparison principle for impulsive systems and some matrix calculation techniques, sufficient conditions for the achievement of quasi-synchronization of coupled complex neural networks are obtained.

Chapter 7: This chapter is devoted to investigating the exponentially cluster synchronization of a class of nonlinearly and nonidentically coupled Lur'e networks with multiple time-varying delays. A novel impulsive pinning control strategy is introduced, which are imposed on the Lur'e systems which have directed connections with any other clusters. Based on the Lyapunov stability theorem, mathematical induction method and the average impulsive interval, the conditions for successful cluster synchronization of Lur'e networks are derived.

Chapter 8: This chapter mainly studies the leader-following synchronization issue for a class of Lur'e networks with nonlinear couplings and multi-delay with various sizes. A kind of impulsive pinning controllers is designed for achieving exponential synchronization. By utilizing the definition of average impulsive interval, parameter variation method and contradiction method, sufficient synchronization criteria are derived. Noticeably, convergence rates of impulses with different functions of impulsive effects are discussed specifically.

Wuxi, China — Ze Tang
Wuxi, China — Dong Ding
Wuxi, China — Yan Wang
Wuxi, China — Zhicheng Ji
Gyeongsan, Republic of Korea — Ju H. Park
July 2021

Acknowledgements

This book is supported by the National Natural Science Foundation of China with Grant No. 61803180, the Natural Science Foundation of Jiangsu Province with Grant No. BK20180599, the China Postdoctoral Science Foundation funded Project with Grant No. 2021T140280, 2020M681484, the Postdoctoral Science Foundation of Jiangsu Province with Grant No. 2021K408C, the 111 Project with Grant No. B12018, the National Key Research and Development Program of China with Grant No. 2018YFB1701903 and the National Research Foundation of Korea with Grant No. 2019R1A5A808029011.

I'd like to express my appreciation to my postgraduate students Miss Yue Gao, Mr. Deli Xuan, Mr. Chenhui Jiang, Mr. Jiafeng Wang and Mr. Kunpeng Wang who have unselfishly given their valuable time in arranging these raw materials into something I'm proud of. Last but not the least, I want to give the most special appreciation to Prof. Jianwen Feng from Shenzhen University, Shenzhen, China for his greatly and kindly supporting to me.

Contents

Symbols

N	1, 2, 3, ...
$\mathbb{R}$	Field of real numbers
$\mathbb{R}^n$	Field of n-dimensional real vector space
$\mathbb{R}^{m\times n}$	Field of $m \times n$ real matrices space
L^T	The Transpose of matrix L
L^{-1}	The inverse of matrix L
$\otimes$	Kronecker product of two matrices
$*$	The symmetrical part in a matrix
$diag\{\cdots\}$	Block-diagonal matrix
$\|\cdot\|$	Euclid norm of the matrix or the vector
$\lvert\cdot\rvert$	Absolute value
$\lambda_{\max}(L)$	The largest eigenvalue of matrix L
$\lambda_{\min}(L)$	The smallest eigenvalue of matrix L
$\max_{1\leq i\leq N}\{\cdot\}$	The maximum value
$\min_{1\leq i\leq N}\{\cdot\}$	The minimum value
$L > 0$	The matrix L is positive definite
$L \geq 0$	The matrix L is positive semi-definite
$I_{n\times n}$	The $n \times n$ real identity matrix
$D^+u(t)$	$D^+u(t) = \overline{\lim}_{h\to 0^+} \frac{u(t+h)-u(t)}{h}$
$u(t_k^+)$	$u(t_k^+) = \lim_{h\to 0^+} u(t_k + h)$
$u(t_k^-)$	$u(t_k^-) = \lim_{h\to 0^-} u(t_k + h)$
sup	Supremum
inf	Infimum

Acronyms

CDNs	Complex dynamical networks
CNNs	Coupled neural networks
Lur'e DNs	Lur'e dynamical networks
NNs	Neural networks

Chapter 1
Introduction

1.1 Background

In the concept of network theory, a complex dynamical network is a connected graph with nontrivial topological characteristics, like high cluster effects, the different degree distributions of each node, homogeneous and heterogeneous among different nodes, different types of hierarchical structures and different community structures. These characteristics do not always appear in simple graphs like random graphs and lattices, but often appear in some graphs when modelling the real systems. As one of the most important collective behaviors, the investigation on synchronization of the complex networks has been lasting for several decades until now. In fact, the synchronization of complex networks has been applied in many fields, such as medical treatment, parallel image grabbing and processing, scientific index networks, smart grid and so on. Specially, due to the importance not only in the theoretical analysis but also in practical applications, the study on impulsive synchronization of complex dynamical networks have attracted engineers and scientists from various disciplines, such as electrical engineering, mechanical engineering, mathematics, network science, system engineering.

Complex dynamical networks have been widely studied by researchers from different fields in the past two decades. Initially, it was discovered that the states of certain systems would have abrupt changes when facing complicated working conditions. Under such a circumstance, impulses are introduced to describe the special phenomenon. Since impulses can be regarded as disturbances with some conditions, the circumstances under that the impulses will have a beneficial effect on synchronization have attracted scholars to study. In Lu et al.'s work [1], the definition of desynchronizing impulses and synchronizing impulses was standardized for the first time, and the corresponding ranges of these two functions of impulsive effects were discussed. In the case of synchronizing impulses [2], Zhao et al. designed a single impulsive controller and successfully achieved exponential synchronization. In [3], the continuous feedback controller was utilized to overcome the desynchronizing

Z. Tang et al., *Impulsive Synchronization of Complex Dynamical Networks*,
https://doi.org/10.1007/978-981-16-5383-4_1

impulses to achieve synchronization. However, it is impractical to deploy controllers globally for all the nodes in a complex network. Therefore, it is particularly critical to give an impulsive pinning control scheme. Notably, for some special network synchronization goals such as cluster synchronization, how to determine the controlled nodes is also a key issue. Designing a control law that can balance control costs and control effects is the core problem we hope to solve.

Different from the analyzing procedures for the continuous systems, the analysis for complex networks with impulsive controllers not only needs to build the relationship between the Lyapunov function and the error vectors of each node, but also needs to discuss the relationship between left and right limitations of the Lyapunov function at the impulse instants. As a result, the selection of the Lyapunov function will greatly affect the complexity of the calculation. Initially, a simple Lyapunov function was constructed to obtain the synchronization conditions of the complex network with nonlinear coupling [4]. In order to solve the synchronization issue for complex networks with time-varying delays, in the work of [5], the integral terms and multiple integral terms related to the delay sizes are introduced in the Lyapunov function. For various complex network models, different types of Lyapunov functions are proposed. However, as the Lyapunov function continues to become more and more complicated, the mathematical proof steps becomes more and more cumbersome. Through the work in [6], an impulsive synchronization theorem was summarized. As long as the conditions of the theorem are satisfied, the synchronization conclusion could be obtained. However, the obtained theorem was a kind of delay-independent theorems, and the model was limited to complex networks with time-varying delays. On the other hand, even though the less conservative theorems can be achieved by constructing some more complicated Lyapunov functions, the difficulties of analyzing the relationship between left and right limits will be enhanced. To tackle such an issue, the parameter variation method is proposed to solve the above contradiction between computational complexity and conservativeness of the theorem. By establishing a more appropriate comparison system, only a simple Lyapunov function is needed to obtain the synchronization conditions, and the exponential synchronization convergence can be accurately estimated. Further, with the in-depth studies of complex network coupling types, the comparison system built in such as [7] has been unable to meet the demand for complex models, which also motivates us to carry out this work.

Based on all discussed above, in this book, we are committed to analyzing the following issues:

1. Since the complex network model is undoubtedly the key point to the synchronization issue, this book will analyze a variety of models including CNs with proportional delays, derivative coupled CNs, CNNs with cluster-tree topology. We will show how parameter variation method is extended and appropriate Lyapunov functions are selected.
2. The impulsive control schemes are applied to synchronize CNs. In consideration of control cost and control efficiency, impulsive controllers will cooperate with pinning negative controllers in some situations. Particularly, for those network

structures with clustering topology, the pinning strategy of impulsive controllers will be introduced.

3. Synchronization theorems in regard to synchronizing impulses and desynchronizing are given. In addition, the exponential convergence rates with respect to different functions of impulsive effects are precisely estimated.

In next section, we will briefly introduce the content of each chapter.

1.2 Book Organization

Chapter 2 aims to study the cluster synchronization for a kind of dynamical complex networks containing nonidentical nonlinear Lur'e systems and proportional delay, which has the features of unbounded and time-varying. Based on the topological structure of nonidentically coupled Lur'e network, an effective impulsive pinning controller is proposed and placed on the Lur'e system with directional path to other clusters. Considering different functions that the impulsive effects play, some sufficient criteria for the cluster synchronization of the nonidentically coupled Lur'e dynamical networks are derived by applying the extended parameters variation formula, the impulsive comparison principle and the concept of average impulsive interval.

Chapter 3 discusses the global exponential synchronization for a class of delay derivative coupled neural networks with multiple time-varying delays and stochastic disturbance. To broaden the fields of synchronization applications in network science, consider cluster-tree topology structure of the coupled neural networks, a novel impulsive pinning control strategy is proposed, which skillfully considered the neural networks in current cluster that directly linked to the neural networks in other clusters.

Chapter 4 studies the adaptive control and the exponential synchronization problem of a kind of derivative coupled complex dynamical networks with proportional delay. Sufficient criteria of the exponential synchronization on complex networks are given based on impulsive control and adaptive pinning control strategies jointly applying the proportional delayed impulsive comparison principle, the extended parameters variation formula and the definition of average impulsive interval.

Chapter 5 concentrates on the exponential synchronization of nonlinear coupled Lur'e networks with proportional delay. For the sake of the heterogeneities existed in different Lur'e systems, the investigation on the global quasi-synchronization instead of complete synchronization for the coupled Lur'e networks is given. Comparing to previous general time delay, as a type of unbounded time-varying delays, the proportional delay largely increases the challenge on network synchronization.

Chapter 6 is mainly focused on the global and exponential synchronization problem for a class of complex dynamical networks with nonidentically coupling and time-varying delays. Since the mismatched parameters in different systems, the quasi-synchronization issue instead of complete synchronization is investigated based on the application of impulsive control schemes. In view of the concept of average impulsive interval, the extended comparison principle for impulsive systems and some matrix calculation techniques, sufficient conditions for the achievement of quasi-synchronization of coupled complex neural networks are obtained.

Chapter 7 is devoted to investigating the exponentially cluster synchronization of a class of nonlinearly and nonidentically coupled Lur'e networks with multiple time-varying delays. A novel impulsive pinning control strategy is introduced, which are imposed on the Lur'e systems which have directed connections with any other clusters. Based on the Lyapunov stability theorem, mathematical induction method and the average impulsive interval, the conditions for successful cluster synchronization of Lur'e networks are derived.

Chapter 8 studies the leader-following synchronization issue for a class of Lur'e networks with nonlinear couplings and multi-delay with various sizes. A kind of impulsive pinning controllers is designed for achieving exponential synchronization. By utilizing the definition of average impulsive interval, parameter variation method, and contradiction method, sufficient synchronization criteria are derived. Noticeably, convergence rates of impulses with different functions of impulsive effects are discussed specifically.

1.3 Preliminaries

Definition 1.1 ([8]) The dynamic system $\dot{x}(t) = f(x(t))$ is said to be mean square stable if for any $\varepsilon > 0$, there is a $\rho(\varepsilon) > 0$ such that $\mathbb{E}\{\|x(t)\|^2\} < \varepsilon,\ \ t > 0$ when $\mathbb{E}\{\|x(0)\|^2\} < \rho(\varepsilon)$. In addition, if $\lim_{t\to\infty} \mathbb{E}\{\|x(t)\|^2\} = 0$, for any initial conditions, then the system is said to be globally mean square asymptotically stable.

Definition 1.2 ([9]) Consider a complex dynamical network

$$\dot{x}_i(t) = f(x_i(t)) + c\sum_{j=1}^{N} a_{ij}\Gamma x_j(t), i = 1, 2, \ldots, N, \tag{1.1}$$

where $x_i(t) = (x_i^1(t), x_i^2(t), \ldots, x_i^n(t))^T$ is the state vector of the i-th system, c is the coupling strength, $A = [a_{ij}] \in \mathbb{R}^{N\times N}$ is the coupling matrix, Γ is the inner connected matrix. It is said to achieve asymptotic synchronization if

$$x_1(t) = x_2(t) = \cdots = x_N(t) = s(t), t \to \infty,$$

where $s(t)$ is a solution of the local dynamics of an isolated node satisfying $\dot{s}(t) = f(s(t))$.

Definition 1.3 ([10]) Consider a complex dynamical network (1.1). Synchronization manifold is defined as follows

$$M = \{[x_1^T(t), x_2^T(t), \ldots, x_N^T(t)]^T \in \mathbb{R}^{Nn\times 1} : x_i(t) = x_j(t), i, j = 1, 2, \ldots, N\}.$$

Definition 1.4 ([11]) A network with N oscillators is said to be cluster synchronized if it satisfies

$$\lim_{t\to\infty} \|x_i(t) - s_{\mu_i}(t)\| = 0, i = 1, 2, \ldots, N,$$

where $s_{\mu_i}(t) \in \mathbb{R}^n$ is a solution of an isolate node and satisfies

$$\dot{s}_{\mu_i}(t) = f(s_{\mu_i}(t)), i = 1, 2, \ldots, N,$$

which describes the identical local dynamics for the nodes in the μ_ith cluster.

Definition 1.5 ([12]) Let N systems as

$$\dot{x}_i(t) = f_i(x_i, t), \ \ i = 1, \ldots, N,$$

then, these systems are said to achieve the asymptotical inner synchronization, if

$$x_1(t) = x_2(t) = \cdots = x_N(t) = s(t), as\ t \to \infty,$$

where $s(t) \in \mathbb{R}^n$ is a solution of a target node.

Definition 1.6 ([13]) Function class QUAD(Δ, P, η): Let $\Delta = \mathrm{diag}\{\delta_1, \ldots, \delta_n\}$ be a diagonal matrix and $P = \mathrm{diag}\{p_1, \ldots, p_n\}$ be a positive-definite diagonal matrix. QUAD(Δ, P, η) denotes a class of continuous functions $f(x, t) : \mathbb{R}^n \times [0, +\infty) \to \mathbb{R}^n$ satisfying

$$(x - y)^T P(f(x, t) - f(y, t) - \Delta(x - y)) \le -\eta(x - y)^T(x - y)$$

for some $\eta > 0$, all $x, y \in \mathbb{R}^n$ and all $t \ge 0$.

Lemma 1.1 ([14] Schur complements) *Given constant symmetric matrices A_1, A_2, A_3, where $A_1 = A_1^T$ and $0 < A_2 = A_2^T$, then $A_1 + A_3^T A_2^{-1} A_3 < 0$ if and only if*

$$\begin{pmatrix} A_1 & A_3^T \\ A_3 & -A_2 \end{pmatrix} < 0\ or \begin{pmatrix} -A_2 & A_3 \\ A_3^T & A_1 \end{pmatrix} < 0.$$

Lemma 1.2 ([15]) *Let $\otimes$ denotes the notation of Kronecker product. Then for a constant α and matrices A, B, C, D with appropriate dimensions, the following properties are easily established:*

1. $(\alpha A) \otimes B = A \otimes (\alpha B)$;
2. $(A + B) \otimes C = A \otimes C + B \otimes C$;
3. $(A \otimes B)(C \otimes D) = (AC) \otimes (BD)$;
4. $(A \otimes B)^T = A^T \otimes B^T$.

Lemma 1.3 ([16] Finsler's lemma) *Let $\zeta \in \mathbb{R}^n$, $\Phi = \Phi^T \in \mathbb{R}^{n\times n}$, $B^{\perp}$ represents a basis for the null-space of B, and $B \in \mathbb{R}^{m\times n}$ such that $rank(B) < n$. The following statements are equivalent:*

1. $\zeta^T \Phi \zeta < 0 \;\; \forall B\zeta = 0, \;\; \zeta \neq 0$
2. $B^{\perp^T} \Phi B^{\perp} < 0$
3. $\exists X \in \mathbb{R}^{n\times m} : \Phi + XB + B^T X^T < 0$

Lemma 1.4 ([17] Jensen Inequality) *For any matrix $M > 0$, scalars γ_1 and γ_2 satisfying $\gamma_2 > \gamma_1$, a vector function $x : [\gamma_1, \gamma_2] \to \mathbb{R}^n$ such that the integrations concerned are well defined, then*

$$\left(\int_{\gamma_1}^{\gamma_2} x(s)ds\right)^T M \left(\int_{\gamma_1}^{\gamma_2} x(s)ds\right) \leq (\gamma_2 - \gamma_1) \int_{\gamma_1}^{\gamma_2} x^T(s)Mx(s)ds.$$

Lemma 1.5 ([18] Lyapunov-Krasovskii Stability Theorem) *Consider the delayed differential equation*

$$x(t) = \dot{f}(t, x(t)).$$

Suppose that f is continuous and $f : \mathbb{R} \times \mathbb{C} \to \mathbb{R}^n$ takes $\mathbb{R}\times$ (bounded sets of $\mathbb{C}$) into bounded sets of $\mathbb{R}^n$, and $u, v, w : \mathbb{R}^+ \to \mathbb{R}^+$ are continuous and strictly monotonically nondecreasing functions, $u(s)$, $v(s)$, $w(s)$ are positive for $s > 0$ with $u(0) = v(0) = 0$. If there exists a continuous functional $V : \mathbb{R} \times \mathbb{C} \to \mathbb{R}$ such that

$$u(\|x\|) \leq V(t, x) \leq v(\|x\|),$$

$$\dot{V}(t, x(t, x(t))) \leq -w(\|x(t)\|),$$

where $\dot{V}$ is the derivation of V along the solutions of the above delayed differential equation, then the solution $x = 0$ of this equation is uniformly asymptotically sable.

Lemma 1.6 ([19]) *Assume that x and y are vectors, then for any positive-definite matrix P, the following inequality holds*

$$-2x^T y \leq \inf_{p>0}\{x^T Px + y^T p^{-1} y\}.$$

Lemma 1.7 ([20]) *Let $U = (u_{ij})_{N\times N}$, $P \in \mathbb{R}^{n\times n}$, $x^T = [x_1^T, x_2^T, \ldots, x_N^T]$, $y^T = [y_1^T, y_2^T, \ldots, y_N^T]$, and $x_i, y_i \in \mathbb{R}^n$, $i = 1, \ldots, N$. If $U = U^T$ and each row sum of U is zero, then*

$$x^T(U \otimes P)y = - \sum_{1\leq i<j\leq N} u_{ij}(x_i - x_j)^T P(y_i - y_j).$$

References

1. Lu, J., Ho, D.W.C., Cao, J.: A unified synchronization criterion for impulsive dynamical networks. Automatica. **46**, 1215–1221 (2010)
2. Zhao, Y., Tang, Z., Feng, J.: Single impulsive controller for exponential synchronization of stochastic Lur'e networks with impulsive disturbance. Discrete Dynamics in Nature and Society. **2013**, 864707 (2013)
3. Ding, D., Tang, Z., Wang, Y., Ji, Z.: Synchronization of nonlinearly coupled complex networks: Distributed impulsive method. Chaos, Solitons and Fractals. **133**, 109620 (2020)
4. Li, Y., Wong, K.W., Liao, X.F., Li, C.D.: On impulsive control for synchronization and its application to the nuclear spin generator system. Nonlinear Analysis: Real World Applications. **10**, 1712–1716 (2009)
5. Wang, Z., Zhang, H.: Synchronization stability in complex interconnected neural networks with nonsymmetric coupling. Neurocomputing. **108**, 84–92 (2013)
6. Yang, Z., Xu, D.: Stability analysis of delay neural networks with impulsive effects. IEEE Transactions on Circuits and Systems' II: Express Briefs. **52**, 517–521 (2005)
7. Tang, Z., Park, J.H., Zheng, W.X.: Distributed impulsive synchronization of Lur'e dynamical networks via parameter variation methods. International Journal of Robust and Nonlinear Control. **28**, 1001–1015 (2018)
8. Wang, Z., Wang, Y., Liu, Y.: Global synchronization for discrete-time stochastic complex networks with randomly occurred nonlinearities and mixed time delays. IEEE Transactions on Neural Networks. **21**, 11–25 (2010)
9. Arenas, A., Díaz-Guilera, A., Kurths, J., Moreno, Y., Zhou, C.: Synchronization in complex networks. Physics Reports. **469**, 93–153 (2008)
10. Boccaletti, S., Latora, V., Moreno, Y., Chavez, M., Hwang, D.U.: Complex networks: Structure and dynamics. Physics Reports. **424**, 175–308 (2006)
11. Wang, K., Fu, X., Li, K.: Cluster synchronization in community networks with nonidentical nodes. Chaos. **19**, 023106 (2009)
12. Li, C., Chen, G.: Synchronization in general complex dynamical networks with coupling delays. Physica A: Statistical Mechanics and its Applications. **343**, 263–278 (2004)
13. Chen, T., Liu, X., Lu, W.: Pinning complex networks by a single controller. IEEE Transactions on Circuits and Systems I: Regular Papers. **54**, 1317–1326 (2007)
14. Boyd, S., El Ghaoui, L., Feron, E., Balakrishnan, V.: Linear matrix inequalities in system and control theory. Society for Industrial and Applied Mathematics, Philadelphia (1994)
15. Loan, C.F.V.: The ubiquitous Kronecker product. Journal of Computational and Applied Mathematics. **123**, 85–100 (2000)
16. de Oliveira, M.C., Skelton, R.E.: Stability tests for constrained linear systems. In: Perspectives in robust control. pp. 241–257. Springer, London (2001)
17. Gu, K., Kharitonov, V.L., Chen, J.: On the stability of time-delay systems. Birkhäuser Boston, Cambridge (2003)
18. Hale, J., Lunel, S.M.V.: Introduction to functional differential equations. Springer Science and Business Media, New York (1993)
19. Wu, J., Jiao, L.: Synchronization in complex dynamical networks with nonsymmetric coupling. Physica A. **386**, 513–530 (2007)
20. Wu, C.W., Chua, L.O.: Synchronization in an array of linearly coupled dynamical systems. IEEE Transactions on Circuits and Systems I: Fundamental Theory and Applications. **42**, 430–447 (1995)

Chapter 2
Cluster Synchronization on CDNs with Proportional Delay: Impulsive Effect Method

2.1 Introduction

Complex dynamical networks have been widely studied by researches from different fields in the past two decades [1, 2]. Collective behavior research reflects the performance of all or part of systems in complex networks, which is one of the important branches of complex network research. As one of the representative collective behaviors in complex dynamic networks, synchronization has attracted extensive attention of researchers [3–11]. The purpose of synchronization is to let all or part of the systems in a complex network achieve a desired goal by adjusting some parameters of the system, coupling strength or applying some effective control inputs. Lots of synchronization modes have been studied because of different purposes in practical situations. Among them, cluster synchronization is a unique problem [12–15]. It requires that the network be divided into several subnetworks, which are called clusters, and then achieve the goal of synchronization of each cluster, but there is no requirement for the behavior of the system between different clusters. For instance, Liu and Chen have studied the cluster synchronization for a kind of coupled complex dynamical networks in [16] by devising some simple intermittent pinning control protocols. In addition, based on the pinning control protocol and the general leader-following model, by taking the random disturbances into consideration, Zhou et al. have mainly studied the problem of cluster synchronization for nonlinearly coupled complex dynamical networks with nonidentical systems in [12].

Impulse is a kind of discontinuously signal, which can give an instantaneous motion into the systems. When the impulse is introduced into the system [17–32], both the positive and negative effects of the impulsive should be taken into consideration. That means, the positive impulses can be used as good discontinuous control inputs to provide instantaneous power for the system and save the control cost to a great extent [17, 18, 26–29]. For instance, the leader-follower synchronization problem for a class of heterogeneous dynamic networks was discussed in [27] by introducing a new distributed impulsive control protocol. On the other hand, negative impulses can be regarded as a kind of interferences to the whole network,

Z. Tang et al., *Impulsive Synchronization of Complex Dynamical Networks*,
https://doi.org/10.1007/978-981-16-5383-4_2

which will lead to the instability and even destruction of the network [22, 24, 31]. The problem of the cooperative synchronization of a class of time delay coupled nonlinear multi-agent systems with impulsive disturbances was discussed in [31] by using some important topological structure of graphs and Lyapunov method. However, in the previous research on impulsive control or impulsive interference, only a limited range of impulsive effects were given [17–23]. In order to simulate the actual situation as much as possible, more impulsive effects should be taken seriously.

In order to reflect the actual situation of complex network in practical application and artificial society, it is not enough to only give the coupled network model under general delay state. In some practical cases, when simulating the information transmission between different systems in complex networks, not only the coupling in general state and/or delay state, but also the coupling in proportional delay state should be considered [33–36]. It is well known that neural networks always possess spatial characteristics because there are many parallel pathways with different axon sizes and lengths. Therefore, in a certain time interval, we need to consider a certain expected proportional delay to establish the neural network model. Normally, the term $y(rt)$ is called proportional delay, and $r \in (0, 1)$ is usually described as the proportional delay rate. Taking the proportional delay function into consideration, by using the definition of the delay term $\tau(t) = (1 - r)t$, one could get that when t tends to infinity, $\tau(t)$ tends to infinity. Based on the above discussion, the features of the proportional delay is unbounded, time-varying, and monotonically increasing. In this respect, it is difficult to control the complex networks with proportional delay. The delay function $\tau(t)$ is always assumed to be bounded with $0 \leq \tau(t) \leq \tau$ in previous works [14, 15, 17]. By constructing some effective Lyapunov functions, introducing some inequalities and considering the boundary of time delay, we can analyze the stability of the system, but this method can not be applied to the proportional delay system immediately. It is necessary to study different types of synchronization problems of complex dynamic networks with proportional delay in theory and practice. For example, some finite-time stability problems for the fuzzy cellular neural networks with certain proportional delays is studied in [34] by introducing some useful inequalities and the finite-time stability theories. In addition, by using matrix theory and Lyapunov functional method, the asymptotic stability of a class of cellular neural networks with multiple proportional delays is studied in [36], and its equilibrium point is investigated. But it's worth noting that in most previous works such as [33, 35, 36], a effective equation $y_i(t) = z_i(e^t)$ is applied to study the synchronization of complex networks, which increases the difficulty and complexity on calculations to a great extent.

As far as we know, the problem of cluster synchronization in impulsive control of nonidentical coupled Lur'e networks with proportional delay has not been widely concerned. The importance of this research, both in theory and in practice, urges us to go on this work. In this chapter, we study the cluster synchronization of coupled complex dynamic networks. Different from previous works, this paper considers an unbounded time-varying delay, the proportional delay, to simulate more practical situations. The main contributions of this chapter are as follows: (1) In this chapter, we discuss the problem of cluster synchronization in nonidentically coupled Lur'e networks with unbounded proportional delay. (2) An effective impulsive pinning controller is cleverly designed by thinking about the cluster-tree topology of the Lur'e networks, and they are brought to bear on the systems in the clusters that are directly linked to those in the other clusters. (3) According to the different functions of impulse effects, by using the delayed impulsive comparison principle, the concept of the average impulsive interval and the extend formula of the parameters variation at the same time, some criteria for cluster synchronization of the nonidentically coupled Lur'e dynamical networks are derived. (4) Furthermore, some useful parameter equations are constructed to estimate the convergence rate of exponential cluster synchronization.

The left parts of the chapter are arranged as below. Firstly, the main coupled complex network models of this chapter and some necessary preparations are given in Sect. 2.2. Then in Sect. 2.3, the cluster synchronization of the Lur'e networks with the proportional delay will be studied. In Sect. 2.4, to illustrate the effectiveness of the theoretical results, an example is given. Finally, the conclusion of this chapter will be given in Sect. 2.5.

Notations. H^T stands for the transpose of matrix H. $\mathbb{R}^n$ stands for the n-dimensional Euclidean space. $\mathbb{R}^{n\times n}$ means the set of $n \times n$ real matrices. diag$\{\cdots\}$ denotes a diagonal matrix. The symbol $\|\cdot\|$ denotes the Euclid norm. $\lambda_{\max}(H)$ denotes the maximum eigenvalue of matrix H, and $\max_{1\leq i\leq N}\{\cdot\}$ means selecting the maximum value of the matrix or the vector. If matrix H is positive definite or semi-definite, we denote it as $H > 0 (H \geq 0)$. I_n denotes the identity matrix with n dimension. $H = (h_{ij})$ stands for a $N \times N$ matrix H with elements h_{ij} for $i, j = 1, 2, \ldots, N$. In the chapter, the dimension of vectors and matrices will be clear.

2.2 Model Description and Preliminaries

2.2.1 Network Structure

In order to better expand the content of this chapter, some assumptions about Lur'e network with topology are defined as follow. Take the complex networks with N Lur'e systems into consideration. The total number of the clusters l in the network satisfies the following condition: $N > l \geq 2$. $\phi_i = j$ means the i-th Lur'e system belongs to the j-th cluster. Define the set U_j be all Lur'e systems in the j-th cluster and the set $\bar{U}_j$ be all Lur'e systems in the j-th cluster which have direct connections to the Lur'e systems in other clusters. Through the above discussion, we could conclude that the following equations hold: (1) $U_i \bigcap U_j = \emptyset$, for $i \neq j$ and $i, j = 1, 2, \ldots, l$; (2) $\bigcup_{i=1}^{l} U_i = \{1, 2, \ldots, N\}$.

2.2.2 Problem Formulation

In this chapter, the problem of the nonidentically coupled Lur'e networks with proportional delay will be mainly studied.

$$\dot{z}_i(t) = A_{\phi_i} z_i(t) + B_{\phi_i} f_{\phi_i}(C_{\phi_i} z_i(t)) + c\sum_{j=1}^{N} g_{ij}\Gamma z_j(rt), \tag{2.1}$$

where $z_i(t) = [z_i^1(t), z_i^2(t), \ldots, z_i^n(t)]^T \in \mathbb{R}^n$ denotes the state variable of the i-th Lur'e system; $A_{\phi_i} \in \mathbb{R}^{n\times n}$, $B_{\phi_i} \in \mathbb{R}^{n\times m}$ and $C_{\phi_i} \in \mathbb{R}^{m\times n}$ are all constant matrices; $\Gamma = \mathrm{diag}\{r_1, r_2, \ldots, r_n\} \in \mathbb{R}^{n\times n}$ stands for the inner connection matrix with $r_i \geq 0$ and c denotes the coupling strength; $f_{\phi_i} : \mathbb{R}^m \to \mathbb{R}^m$ belongs to a class of continuously differentiable vector-valued nonlinear functions; Matrix $G = (g_{ij}) \in \mathbb{R}^{N\times N}$ is the coupling matrix that satisfies the diffusive condition, that means, $g_{ii} = -\sum_{j\neq i j=1}^{N} g_{ij} = -\sum_{j\neq i j=1}^{N} g_{ji}$, where $g_{ji} = g_{ij} > 0$ if there has a connection between the i-th Lur'e system and the j-th Lur'e system with $i \neq j$, and $g_{ij} = 0$, otherwise; The factor $r \in (0, 1)$ is a constant that involves the history time. It is worth noting that in the coupled Lur'e network model (2.1), dynamics of the i-th Lur'e system at time t is decided by the states $z_j(t)$ and the states $z_j(rt)$ at history time and rt is proportional to time instant t and the proportionality factor is r. It can be concluded that the factor r is regarded as proportional delay. Therefore, we have $rt = t - \tau(t)$, where $\tau(t) = (1-r)t \geq 0$ and $\tau(t) \to \infty$. It can be seen from the above equation that the proportional delay has the features of unbounded and time-varying. Set matrices $C_{\phi_i} = [c_{\phi_i 1}, c_{\phi_i 2}, \ldots, c_{\phi_i m}]^T$ with $c_{\phi_i j} \in \mathbb{R}^{1\times n}$ for $i = 1, 2, \ldots, N$, $j = 1, 2, \ldots, m$. And we have $C_{\phi_i} z_i(t) = [c_{\phi_i 1} z_i(t), c_{\phi_i 2} z_i(t), \ldots, c_{\phi_i m} z_i(t)]^T$, $f_{\phi_i}(C_{\phi_i} z_i(t)) = [f_{\phi_i}^1(c_{\phi_i 1} z_i(t)), f_{\phi_i}^2(c_{\phi_i 2} z_i(t)), \ldots, f_{\phi_i}^m(c_{\phi_i m} z_i(t))]^T$.

We consider an isolated Lur'e system with different system parameters to achieve the cluster synchronization

$$\dot{s}_{\phi_i}(t) = A_{\phi_i} s_{\phi_i}(t) + B_{\phi_i} f_{\phi_i}(C_{\phi_i} s_{\phi_i}(t)), \tag{2.2}$$

where $s_{\phi_i}(t) = [s_{\phi_i}^1(t), s_{\phi_i}^2(t), \ldots, s_{\phi_i}^n(t)]^T \in \mathbb{R}^n$ for $i = 1, 2, \ldots, N$. Furthermore, for making the Lur'e networks (2.1) and the isolated Lur'e systems (2.2) achieve cluster synchronization, in the ϕ_i-th cluster, the solution $s_{\phi_i}(t)$ could be the leader. While, on the other hand, the Lur'e systems in the ϕ_i-th cluster are followed by the leader $s_{\phi_i}(t)$.

Set $e_i(t) = z_i(t) - s_{\phi_i}(t)$ be the vector of error between $z_i(t)$ and $s_{\phi_i}(t)$ where $e_i(t) = [e_i^1(t), e_i^2(t), \ldots, e_i^n(t)]^T \in \mathbb{R}^n$. Later the error Lur'e networks by adding the control input $u_i(t)$ could be given

$$\dot{e}_i(t) = A_{\phi_i} e_i(t) + B_{\phi_i}(f_{\phi_i}(C_{\phi_i} z_i(t)) - f_{\phi_i}(C_{\phi_i} s_{\phi_i}(t)))$$
$$+ c\sum_{j=1}^{N} g_{ij}\Gamma e_j(rt) + c\sum_{j=1}^{N} g_{ij}\Gamma s_{\phi_j}(rt) + u_i(t), \tag{2.3}$$

where $u_i(t)$ denotes the control input of the controller and it is designed as below

$$u_i(t) = u_{i,0} + \begin{cases} -k_i\Gamma e_i(t) - c\sum_{j=1}^{N} g_{ij}\Gamma s_{\phi_j}(rt), & i \in \bar{U}_{\phi_i}, \\ 0, & i \in U_{\phi_i} - \bar{U}_{\phi_i}, \end{cases} \tag{2.4}$$

where $u_{i,0} = \mu\sum_{k=1}^{\infty} e_i(t)\delta(t - t_k)$, and k_i stands for the control gain, the parameter μ always be regarded as the impulsive effect and it will be investigate in the next content, the function $\delta(\cdot)$ denotes the Dirac function. The sequence of time series $\zeta = \{t_1, t_2, \ldots\}$ is strictly increasing and it denotes the impulsive instants, which satisfies $t_{k-1} < t_k$ and $\lim_{k\to+\infty} t_k = +\infty$ for $k = 1, 2, \ldots$. $K = \text{diag}\{k_1, k_2, \ldots, k_n\}$ means the control gain matrix, which could show the strategy of control.

Base on the above discussion, the following error systems could be obtained by taking the coupled error Lur'e network (2.3) and the negative feedback impulsive pinning controller (2.4) into consideration

$$\begin{cases} \dot{e}_i(t) = A_{\phi_i} e_i(t) + B_{\phi_i}\Phi_{\phi_i}(C_{\phi_i} e_i(t)) - k_i\Gamma e_i(t) \\ \qquad +c\sum_{j=1}^{N} g_{ij}\Gamma e_j(rt), \ t \neq t_k, k = 1, 2, \ldots \\ \Delta e_i(t_k) = \mu e_i(t_k^-), \ k = 1, 2, \ldots, \\ e_i(0) = e_i^0, \ i = 1, 2, \ldots, N, \end{cases} \tag{2.5}$$

where $\Phi_{\phi_i}(C_{\phi_i} e_i(t)) = f_{\phi_i}(C_{\phi_i} z_i(t)) - f_{\phi_i}(C_{\phi_i} s_{\phi_i}(t)))$. Furthermore, in order to facilitate the follow-up study in this chapter, let $e_i(t)$ be right-hand continuous at $t = t_k, k = 1, 2, \ldots$, and $e_i(t_k) = e_i(t_k^+) = \lim_{\epsilon\to 0^+} e_i(t_k + \epsilon), e_i(t_k^-) = \lim_{\epsilon\to 0^-} e_i(t_k + \epsilon)$. Let $e^0 = [e_1^0, e_2^0, \ldots, e_N^0]^T$. Thus, at discontinuous time instants $t = t_k$ for $k = 1, 2, \ldots$, the solutions of the nonidentically coupled error Lur'e networks (2.5) satisfy the condition of piecewise right-hand continuous.

2.2.3 *Preliminaries*

In order to achieve the cluster synchronization on the Lur'e networks, we need to do some necessary preliminaries as below.

Definition 2.1 In this chapter $S_U = \{z(t) = [z_1(t)^T, z_2(t)^T, \ldots, z_N(t)^T]^T | z_i(t) \in \mathbb{R}^n, z_i(t) = z_j(t), \forall i, j \in U_k, \ i, j = 1, 2, \ldots, N, k = 1, 2, \ldots, l\}$ is defined as the synchronization manifold of the nonidentically coupled Lur'e network (2.1) with l clusters.

Definition 2.2 ([16]) If the error vectors satisfies $\lim_{t\to+\infty}\|z_i(t)-z_j(t)\|=0$ when $\phi_i=\phi_j$, and $\lim_{t\to+\infty}\|z_i(t)-z_j(t)\|\neq 0$ for $\phi_i\neq\phi_j$ with $i, j=1,2,\ldots,N$ for all initializations, then S_U that is defined in Definition 2.1 will become globally stable, i.e., the objective of the cluster synchronization between the nonidentically coupled Lur'e networks (2.1) and the isolated Lur'e system (2.2) is achieved.

Definition 2.3 ([17]) Suppose $N_\zeta(\bar{t},t)$ denotes the impulse numbers of the impulsive sequence. $(t,\bar{t})$ and $\zeta=\{t_1,t_2,\ldots\}$ are the time interval and the impulsive sequence, respectively. When $\bar{t}\geq t\geq 0$, if the relationship between the parameter $N_a>0$ and $N_0>0$ satisfy the following inequality

$$\frac{\bar{t}-t}{N_a}-N_0\leq N_\zeta(\bar{t},t)\leq\frac{\bar{t}-t}{N_a}+N_0, \tag{2.6}$$

then one could draw the conclusion that N_a is larger than the average impulsive interval of the impulsive sequence.

Lemma 2.1 ([18]) *Set $\varphi(t)$ is an almost continuously valued function. At time instants such as t_k, $\varphi(t)$ is right-hand continuous, which means $\varphi(t_k^+)=\varphi(t_k)$ and $\varphi(t_k^-)$ denotes the left-hand of the time instant t_k. $PC(l)=\{\varphi|\varphi:[-\tau,\infty)\to\mathbb{R}^l\}$ denotes the set with l order piecewise continuous functions and $0\leq\tau(t)\leq\tau$. If there have two functions $V(t)$ and $\upsilon(t)$ belong to the set $PC(l)$ with $l=1$ and a, b and ς that are all constants satisfy*

$$\begin{cases} D^+V(t)\leq aV(t)+bV(t-\tau(t)),\ t\neq t_k, k\in\mathcal{N},\\ V(t_k)\leq\varsigma V(t_k^-),\ k\in\mathcal{N},\end{cases}$$

$$\begin{cases} D^+\upsilon(t)>a\upsilon(t)+b\upsilon(t-\tau(t)),\ t\neq t_k, k\in\mathcal{N},\\ \upsilon(t_k)=\varsigma\upsilon(t_k^-),\ k\in\mathcal{N},\end{cases}$$

and when t meets the condition $-\tau\leq t\leq 0$, we could have $V(t)\leq\upsilon(t)$, then for all time $t>0$ one has $V(t)\leq\upsilon(t)$.

Assumption 2.1 Set the nonlinear function $f_{\phi_i}(\cdot)$ satisfy Lipschitz condition, in other words, by introducing some Lipschitz constants $l_{\phi_i}>0$, we have

$$\|f_{\phi_i}(x)-f_{\phi_i}(y)\|\leq l_{\phi_i}\|x-y\|,$$

for any two vectors $x, y\in\mathbb{R}^n$, $i=1,2,\ldots,N$ that are n dimensional.

2.3 Main Results

In this section, by introducing the impulsive pinning controller (2.4), we will discuss the exponential cluster synchronization for a class of nonidentically coupled Lur'e dynamical networks with proportional delay.

Theorem 2.1 *Suppose that by imposing the impulsive pinning controller (2.4), Assumption 2.1 holds in the nonidentically coupled error Lur'e networks (2.3). Consider the impulsive sequence* $\zeta = \{t_1, t_2, \ldots\}$, N_a *is larger than the average impulsive interval. Define* $\varpi = (1+\mu)^2$. *If there are the matrix* $K = diag\{k_1, k_2, \ldots, k_N\}$ *that is non-negative definite and some constants* a, b, l_{ϕ_i} $(i = 1, 2, \ldots, N)$ *that are positive satisfy*

Case I*: Consider the equation of* $\varpi = (1+\mu)^2$, *i.e., for the impulsive effect* μ *satisfying* $0 < \varpi \leq 1$,
(i) *The matrix satisfies*

$$\Psi = \begin{bmatrix} aI_N - 2K & cG \\ * & -bI_N \end{bmatrix} < 0; \tag{2.7}$$

(ii-a) $\delta = \frac{\ln \varpi}{N_a} - p < 0$, *a positive constant* $0 < \lambda < -\delta$ *satisfies*

$$\delta + \lambda r + \xi \varpi^{-N_0} < 0, \tag{2.8}$$

$$\lambda(\delta + \lambda r + r\xi \varpi^{-N_0}) + \delta(\delta + \lambda r + \xi \varpi^{-N_0}) > 0, \tag{2.9}$$

$$\lambda^2(\delta + \lambda r + r\xi \varpi^{-N_0}) > \delta^2(\delta + \lambda r + \xi \varpi^{-N_0}), \tag{2.10}$$

where $p = -\max_{1\leq i\leq N}\{\lambda_{\max}(A_{\phi_i} + A_{\phi_i}^T + B_{\phi_i}B_{\phi_i}^T + l_{\phi_i}^2 C_{\phi_i}^T C_{\phi_i} - a\Gamma)\}$, $\xi = b\max_{1\leq i\leq N}\{r_i\}$, *then the error Lur'e networks (2.5) could achieve the objective of exponentially stable and the convergence rate is* $\frac{\lambda}{2}$. *That is to say, if the impulsive effect satisfying* $-2 \leq \mu \leq 0$ *and* $\mu \neq -1$, *the objective of the cluster synchronization between the nonidentically coupled Lur'e networks (2.1) and the nonlinear Lur'e systems (2.2) is finally achieved by proposing the impulsive pinning controller (2.4).*

Case II*: Consider the equation of* $\varpi = (1+\mu)^2$, *if the impulsive effect* μ *satisfying* $\varpi > 1$, *and if the condition (i) in* ***Case I*** *is satisfied and*
(ii-b) *If there exists a constant* λ *satisfying* $0 < \lambda' < -\delta$ *and there have the following equations*

$$\delta + \lambda' r + \xi \varpi^{N_0} < 0, \tag{2.11}$$

$$\lambda'(\delta + \lambda' r + r\xi \varpi^{N_0}) + \delta(\delta + \lambda' r + \xi \varpi^{N_0}) > 0, \tag{2.12}$$

$$\lambda'^2(\delta + \lambda r + r\xi \varpi^{N_0}) > \delta^2(\delta + \lambda' r + \xi \varpi^{N_0}), \tag{2.13}$$

then the error Lur'e networks (2.5) could achieve the objective of exponentially stable and the convergence rate is $\frac{1}{2}\lambda'$. *In other words, if the impulsive effect satisfying* $\mu <$

-2 *or* $\mu > 0$, *the objective of the cluster synchronization between the nonidentically coupled Lur'e networks with proportional delay (2.1) and the target nonlinear Lur'e systems (2.2) would finally achieved after adding the impulsive pinning controller (2.4).*

Proof Construct the Lyapunov function as below

$$V(t) = \sum_{i=1}^{N} e_i^T(t)e_i(t). \tag{2.14}$$

When t reaches the impulsive instants such as $t = t_k, k = 1, 2, \ldots$, based on the impulsively controlled error Lur'e network (2.5) we could derive

$$\begin{aligned} V(t_k^+) &= \sum_{i=1}^{N} e_i^T(t_k^+)e_i(t_k^+) \\ &= \sum_{i=1}^{N}(1+\mu)^2 e_i^T(t_k^-)e_i(t_k^-) = \varpi V(t_k^-). \end{aligned} \tag{2.15}$$

Then, suppose $t \in [t_{k-1}, t_k)$ with $k = 1, 2, \ldots$, calculating the derivative values of (2.14) along the impulsively controlled Lur'e network (2.5) gives

$$\begin{aligned} \dot{V}(t) &= 2\sum_{i=1}^{N} e_i^T(t)(A_{\phi_i}e_i(t) + B_{\phi_i}\Phi_{\phi_i}(C_{\phi_i}e_i(t)) \\ &\quad - k_i\Gamma e_i(t) + c\sum_{j=1}^{N} g_{ij}\Gamma e_j(rt)) \\ &\le \sum_{i=1}^{N} e_i^T(t)(A_{\phi_i} + A_{\phi_i}^T + B_{\phi_i}B_{\phi_i}^T + l_{\phi_i}^2 C_{\phi_i}^T C_{\phi_i} \\ &\quad - 2k_i\Gamma)e_i(t) + 2c\sum_{i=1}^{N}\sum_{j=1}^{N} g_{ij}e_i^T(t)\Gamma e_j(rt) \\ &= \sum_{i=1}^{N} e_i^T(t)(2A_{\phi_i} + B_{\phi_i}B_{\phi_i}^T + l_{\phi_i}^2 C_{\phi_i}^T C_{\phi_i} - a\Gamma)e_i(t) \\ &\quad + b\sum_{i=1}^{N} e_i^T(rt)\Gamma e_i(rt) - b\sum_{k=1}^{n} r_k\tilde{e}^k(rt)^T\tilde{e}^k(rt) \\ &\quad + a\sum_{k=1}^{n} r_k\tilde{e}^k(t)^T\tilde{e}^k(t) - 2\sum_{k=1}^{n} r_k\tilde{e}^k(t)^T K\tilde{e}^k(t) \end{aligned}$$

$$
\begin{aligned}
&+ 2c\sum_{k=1}^{n} r_k \tilde{e}^k(t)^T G\tilde{e}^k(rt) \\
&= \sum_{i=1}^{N} e_i^T(t)(2A_{\phi_i} + B_{\phi_i}B_{\phi_i}^T + l_{\phi_i}^2 C_{\phi_i}^T C_{\phi_i} - a\Gamma)e_i(t) \\
&+ b\sum_{i=1}^{N} e_i^T(rt)\Gamma e_i(rt) + \sum_{k=1}^{n} r_k\eta^T(t)\Psi\eta(t) \\
&\leq -pV(t) + \xi V(rt),
\end{aligned}
\tag{2.16}
$$

where the vector $\eta(t) = [\tilde{e}^k(t)^T, \tilde{e}^k(rt)^T]^T$ and $\tilde{e}^k(t) = [e_1^k(t), e_2^k(t), \ldots, e_N^k(t)]^T$ for $k = 1, 2, \ldots, n$.

According to the equation (2.15) and the inequality (2.16), based on the comparison principle, the following impulsive comparison systems with proportional delay could be derived, which has and only has one solution $\xi(t)$ for any $\varepsilon > 0$.

$$
\begin{cases}
\dot{v}(t) = -pv(t) + \xi v(rt) + \varepsilon, \ t \neq t_k, k = 1, 2, \ldots \\
v(t_k^+) = \varpi v(t_k^-), \ t = t_k, k = 1, 2, \ldots \\
v(0) = \|e^0\|^2 = \sum_{i=1}^{N} \|e_i^0\|^2.
\end{cases}
\tag{2.17}
$$

When $t > 0$, one could conclude that for Lyapunov function $V(t)$, we have $V(t) \leq v(t)$. By the improved parameters variation formula [20], the following integral equation of the impulsive comparison system is:

$$
v(t) = M(t, 0) \cdot v(0) + \int_0^t M(t, s)(\xi v(rs) + \varepsilon)ds,
\tag{2.18}
$$

where $M(t, s)$ is the Cauchy matrix of the following linear impulsive system:

$$
\begin{cases}
\dot{v}(t) = -pv(t), \quad t \neq t_k, \\
v(t_k^+) = \varpi \cdot v(t_k^-),
\end{cases}
$$

and additional, the basic form of $M(t, s)$ is

$$
\begin{aligned}
M(t, s) &= e^{-p(t-s)} \prod_{s \leq t_k \leq t} \varpi \\
&\leq e^{-p(t-s)} \varpi^{N_\zeta(t,s)}.
\end{aligned}
$$

In the next content, according to different roles that the impulsive effect μ played in the cluster synchronization, the range of ϖ is divided into two cases to improve preciseness.

(Case I). If $0 < \varpi \leq 1$, by using the concept of average impulsive interval $N_\zeta(t,s)$, the Cauchy matrix $M(t,s)$ of the linear impulsive system has the following form:

$$\begin{aligned} M(t,s) &= \leq e^{-p(t-s)} \varpi^{N_\zeta(t,s)} \\ &\leq e^{-p(t-s)} \varpi^{\frac{t-s}{N_a} - N_0} \\ &\leq \varpi^{-N_0} e^{(\frac{\ln \varpi}{N_a} - p)(t-s)}. \end{aligned} \tag{2.19}$$

Combining the Eq. (2.18) and the Cauchy matrix of the linear impulsive system (2.19), we have

$$\begin{aligned} v(t) &= M(t,0) \cdot v(0) + \int_0^t M(t,s)(\xi v(rs) + \varepsilon) ds \\ &\leq \varpi^{-N_0} \sum_{i=1}^{N} \|e_i^0\|^2 e^{(\frac{\ln \varpi}{N_a} - p)t} + \int_0^t \varpi^{-N_0} e^{(\frac{\ln \varpi}{N_a} - p)(t-s)} (\xi v(rs) + \varepsilon) ds \\ &= \varpi_0 e^{\delta t} + \int_0^t e^{\delta(t-s)} \varpi^{-N_0} (\xi v(rs) + \varepsilon) ds, \end{aligned} \tag{2.20}$$

where $\varpi_0 = \varpi^{-N_0} \sum_{i=1}^{N} \|e_i^0\|^2$ and $\delta = \frac{\ln \varpi}{N_a} - p$.

Next, on account of the inequality (2.20), using the proof by contradiction method in math, we could derive some useful inequalities to evaluate the upper bound of $v(t)$. In this connection, it is essential that the parameter δ satisfies $\delta < 0$.

According to the above deduction and research, for $\forall t \geq 0$ and the given λ satisfies (2.8)–(2.10), we need to prove that there holds

$$v(t) < \varpi_0 e^{-\lambda t} - \frac{\varpi^{-N_0} \varepsilon}{(\frac{\ln \varpi}{N_a} - p) + \xi \varpi^{-N_0}}. \tag{2.21}$$

It is particular that one has

$$v(0) < \varpi_0 - \frac{\varpi^{-N_0} \varepsilon}{(\frac{\ln \varpi}{N_a} - p) + \xi \varpi^{-N_0}}.$$

In the next, by introducing the proof by contradiction method in math, we will prove that the inequality (2.21) holds. If $t > 0$, the inequality (2.21) does not hold, that is to say, there must be a positive time instant $\bar{t}$ to make the following inequality true

$$v(\bar{t}) \geq \varpi_0 e^{-\lambda \bar{t}} - \frac{\varpi^{-N_0} \varepsilon}{(\frac{\ln \varpi}{N_a} - p) + \xi \varpi^{-N_0}}, \ \bar{t} > t > 0, \tag{2.22}$$

however, the results in assumption inequality (2.21) is still satisfied for $0 < t < \bar{t}$. Hence, based on (2.20) and (2.21), (2.22), we have the following derivation where

$\theta = -\frac{\varpi^{-N_0}\varepsilon}{\delta+\xi\varpi^{-N_0}}$

$$
\begin{aligned}
v(\bar{t}) &\le \varpi_0 e^{\delta\bar{t}} + \int_0^{\bar{t}} e^{\delta(\bar{t}-s)}\varpi^{-N_0}(\xi v(rs)+\varepsilon)ds \\
&< \varpi_0 e^{\delta\bar{t}} + \int_0^{\bar{t}} e^{\delta(\bar{t}-s)}\xi\varpi^{-N_0}(\varpi_0 e^{-\lambda rs} \\
&\quad - \frac{\varpi^{-N_0}\varepsilon}{\delta+\xi\varpi^{-N_0}})ds + \int_0^{\bar{t}} e^{\delta(\bar{t}-s)}\varpi^{-N_0}\varepsilon ds \\
&= \varpi_0 e^{\delta\bar{t}} + \xi\varpi^{-N_0}\varpi_0 e^{\delta\bar{t}} \int_0^{\bar{t}} e^{-(\delta+\lambda r)s}ds \\
&\quad + (\xi\varpi^{-N_0}\theta + \varpi^{-N_0}\varepsilon)\int_0^{\bar{t}} e^{\delta(\bar{t}-s)}ds \\
&= \varpi_0 e^{\delta\bar{t}} - \frac{\xi\varpi_0\varpi^{-N_0}}{\delta+\lambda r}(e^{-\lambda r\bar{t}} - e^{\delta\bar{t}}) \\
&\quad + \frac{\varpi^{-N_0}(\xi\theta+\varepsilon)}{-\delta}(1-e^{\delta\bar{t}}) \\
&\le \varpi_0 e^{\delta\bar{t}} - \frac{\xi\varpi_0\varpi^{-N_0}}{\delta+\lambda r}(e^{-\lambda r\bar{t}} - e^{\delta\bar{t}}) + \theta(1-e^{\delta\bar{t}}).
\end{aligned} \tag{2.23}
$$

To better use the proof by contradiction method in math from (2.23), we need to construct some necessary equations. The first parameter function is described as

$$h(t) = \delta(\delta+\lambda r+\xi\varpi^{-N_0})e^{\delta t} + \lambda(\delta+\lambda r+r\xi\varpi^{-N_0})e^{-\lambda t}. \tag{2.24}$$

The derivative value of $h(t)$ is

$$\dot{h}(t) = \delta^2(\delta+\lambda r+\xi\varpi^{-N_0})e^{\delta t} - \lambda^2(\delta+\lambda r+r\xi\varpi^{-N_0})e^{-\lambda t}.$$

In particular, we have $\dot{h}(t) = 0$ if and only if t satisfies

$$t = N_h = \frac{1}{\delta+\lambda}\ln\frac{\lambda^2((\delta+\lambda r)+r\xi\varpi^{-N_0})}{\delta^2((\delta+\lambda r)+\xi\varpi^{-N_0})},$$

where

$$0 < \frac{\lambda^2((\delta+\lambda r)+r\xi\varpi^{-N_0})}{\delta^2((\delta+\lambda r)+\xi\varpi^{-N_0})} < 1$$

according to the condition (2.10) proposed from Theorem 2.1. Furthermore, for $t \in (0, N_h)$ one have $\dot{h}(t) > 0$ and it means that $h(t)$ is an increasing function in the section of $(0, N_h)$. On the other hand, for $t \in (N_h, +\infty)$ we could have $\dot{h}(t) < 0$ and it implies that $h(t)$ is a decreasing function in the section of $t \in (N_h, +\infty)$. Moreover, we have

$$h(0) = \lambda[(\delta + \lambda r) + r\xi\varpi^{-N_0}] + \delta[(\delta + \lambda r) + \xi\varpi^{-N_0}] > 0$$

by condition (2.9) in Theorem 2.1. And according to (2.24), we could derive

$$\lim_{t \to +\infty} h(t) = 0$$

with $\delta < 0$. Hence, when $t \geq 0$, $h(t) \geq 0$.

From condition (2.8) in Theorem 2.1, we could obtain $\delta + \lambda r < 0$. And because of the existence of $h(t) \geq 0$, we have

$$\delta(1 + \frac{\xi\varpi^{-N_0}}{\delta + \lambda r})e^{\delta t} + \lambda(1 + \frac{r\xi\varpi^{-N_0}}{\delta + \lambda r})e^{-\lambda t} \leq 0. \tag{2.25}$$

Since the proportional delay rate $r \in (0, 1)$ and $\lambda \in (0, -\delta)$, we could get $e^{-\lambda r t} > e^{-\lambda t}$. Then define the second parameter equation

$$\bar{h}(t) \triangleq \delta(1 + \frac{\xi\varpi^{-N_0}}{\delta + \lambda r})e^{\delta t} + \lambda\frac{r\xi\varpi^{-N_0}}{\delta + \lambda r}e^{-\lambda r t} + \lambda e^{-\lambda t}.$$

And it can be easily proved that $\bar{h}(t) \leq 0$ from (2.25). Define the parameter function $H(t)$ as

$$H(t) = (1 + \frac{\xi\varpi^{-N_0}}{\xi + \lambda r})\, e^{\delta t} - \frac{\xi\varpi^{-N_0}}{\xi + \lambda r}e^{-\lambda r t} - e^{-\lambda t}.$$

From the above equation, it could be verified the existence of $H(0) = 0$ and $\dot{H}(t) = \bar{h}(t)$, which means

$$\dot{H}(t) = \bar{h}(t) \leq 0.$$

Therefore, one obtains that $H(t)$ is a monotonically decreasing function and $H(0) = 0$. Most importantly, one could conclude that for all $t \geq 0$, there has

$$H(t) \leq H(0) = 0,$$

in other words,

$$(1 + \frac{\xi\varpi^{-N_0}}{\delta + \lambda r})\, e^{\delta t} - \frac{\xi\varpi^{-N_0}}{\delta + \lambda r}e^{-\lambda r t} \leq e^{-\lambda t}, \qquad t \geq 0. \tag{2.26}$$

Based on the existence of inequalities (2.23) and (2.26), one obtains

$$
\begin{aligned}
v(\bar{t}) &\leq \varpi_0 e^{\delta \bar{t}} - \frac{\xi \varpi_0 \varpi^{-N_0}}{\delta + \lambda r}(e^{-\lambda r \bar{t}} - e^{\delta \bar{t}}) + \theta(1 - e^{\delta \bar{t}}) \\
&= \varpi_0 (1 + \frac{\xi \varpi^{-N_0}}{\delta + \lambda r}) e^{\delta \bar{t}} - \varpi_0 \frac{\xi \varpi^{-N_0}}{\delta + \lambda r} e^{-\lambda r \bar{t}} + \theta(1 - e^{\delta \bar{t}}) \\
&< \varpi_0 \, e^{-\lambda \bar{t}} + \theta(1 - e^{\delta \bar{t}}) \\
&< \varpi_0 \, e^{-\lambda \bar{t}} + \theta \\
&= \varpi_0 e^{-\lambda \bar{t}} - \frac{\varpi^{-N_0} \varepsilon}{\delta + \xi \varpi^{-N_0}},
\end{aligned} \tag{2.27}
$$

which is contradictory with the assumption (2.22). Then we could proof that (2.22) is wrong. That is to say, for $t \geq 0$, the assumption (2.21) holds. Based on the impulsive comparison principle, one has

$$
0 < V(t) \leq v(t) < \varpi_0 e^{-\lambda t} - \frac{\varpi^{-N_0} \varepsilon}{\delta + \xi \varpi^{-N_0}}.
$$

when $\varepsilon \to 0$, we have

$$
\|e(t)\| \leq \sqrt{\varpi^{-N_0} \sum_{i=1}^{N} \|e_i^0\| \, e^{-\frac{\lambda}{2} t}}.
$$

Moreover, one has

$$
\lim_{t \to +\infty} \|e(t)\| = 0,
$$

which indicates that the objective of exponentially stable of the nonidentically coupled error Lur'e networks (2.5) could be achieved. Up to now, for $0 < \varpi \leq 1$, after adding the designed impulsive pinning controller (2.4), the nonidentically coupled nonlinear Lur'e networks with proportional delay (2.1) and the target nonlinear Lur'e systems (2.2) could eventually achieve cluster synchronization and by calculating, the exponential convergence rate is $\frac{\lambda}{2}$.

(Case II). If $\varpi > 1$, then the Cauchy matrix of the linear impulsive system $M(t, s)$, like Case I, can be calculated as

$$
\begin{aligned}
M(t, s) &= e^{-p(t-s)} \varpi^{N_\zeta(t,s)} \\
&\leq e^{-p(t-s)} \varpi^{\frac{t-s}{N_a} + N_0} \\
&\leq \varpi^{N_0} e^{(\frac{\ln \varpi}{N_a} - p)(t-s)}.
\end{aligned} \tag{2.28}
$$

From the inequality (2.20), one obtains

$$
v(t) \leq \bar{\varpi}_0 e^{\delta t} + \int_0^t e^{\delta(t-s)} \varpi^{N_0} (\xi v(rs) + \varepsilon) ds, \tag{2.29}
$$

where $\bar{\varpi}_0 = \varpi^{N_0} \sum_{i=1}^{N} \|e_i^0\|^2$.

Similarly, from the conditions (2.11)–(2.13) in Theorem 2.1 with $\bar{\lambda} \in (0, -\delta)$, we could verify that for $\forall t \geq 0$, there holds

$$v(t) < \bar{\varpi}_0 e^{-\bar{\lambda} t} - \frac{\varpi^{N_0}\varepsilon}{\delta + \xi \varpi^{N_0}}.$$

And then we also have

$$0 < V(t) \leq v(t) < \bar{\varpi}_0 e^{-\bar{\lambda} t} - \frac{\varpi^{N_0}\varepsilon}{\delta + \xi \varpi^{N_0}}.$$

Letting $\varepsilon \to 0$ gives

$$\|e(t)\| \leq \sqrt{\varpi^{N_0} \sum_{i=1}^{N} \|e_i^0\|\, e^{-\frac{\bar{\lambda}}{2} t}}.$$

Further, we have

$$\lim_{t \to +\infty} \|e(t)\| = 0.$$

With similar derivation described before, in the end, the nonidentically coupled error Lur'e networks (2.5) could achieve exponential stable and the convergence rate is $\frac{\bar{\lambda}}{2}$. That is to say, for $\varpi > 1$, after adding the designed impulsive pinning controller (2.4), the nonidentically coupled nonlinear Lur'e networks with proportional delay (2.1) and the target nonlinear Lur'e systems (2.2) could eventually achieve cluster synchronization. Based on the above discussion, we finally finish the proof of Theorem 2.1 by taking two different cases into consideration. □

Remark 2.1 Generally, there are some typical methods to study time-varying delay or impulsive systems with time invariance. Firstly, in [21, 22], Halanay's inequality is introduced. By using this inequality, for $t \in [t - \tau, t]$, one could have the impulse differential inequality $D^+V(t) \leq -pV(t) + r \sup_{s \in [t-\tau, t]}\{V(s)\}$ with initial values $V(t) = \phi(t)$, and then for $t \geq t_0$ we could derive $V(t) \leq \sup_{s \in [t-\tau, t]}\{V(s)\}e^{-\lambda(t-t_0)}$, and λ is the unique solution to the equation $\lambda - p + re^{\lambda t} = 0$. Secondly, the upper bounded of the comparison vector described before could be estimated by using the method of variation of parameters [17, 20, 24, 27]. Thirdly, the upper bound of the error vector could be induced well in the form of $p^k V(t_0)e^{-\lambda(t-t_0)}$ by applying the method of mathematical induction [23]. It is worth noting that all methods must meet the condition that the time-varying delay $\tau(t)$ should satisfies the range of $0 \leq \tau(t) \leq \tau$. However, in this chapter, the model of the complex networks contain proportional delay, which has the characteristic of no boundary. With that in mind, the Halanay's inequality and some traditional methods can not be used here. Hence, by cleverly constructing some necessary equations $h(t)$, $\bar{h}(t)$ and $H(t)$, we could use the impulsive comparison principle and the proof by contradiction method in math to estimate the upper bound of the comparison vector $v(t)$ well.

Remark 2.2 In general, considering systems that contain proportional delay, at time instant t, the dynamic state of the i-th Lur'e system depends on both $z_i(t)$ and $z_i(rt)$. As the proportional delay factor, for the state $z_i(t)$ at time t and the historical state $z_i(rt)$ at time rt for $i = 1, 2, \ldots, N$, $r \in (0, 1)$ indicates the ratio of time between them. The proportional delay has the characteristic of no boundary, which can be distinguished from the time invariant delay τ and time-varying delay $\tau(t)$. Because we could easily see that no matter what the value of proportional delay factor $r \in (0, 1)$ is, as t goes to infinity, the term rt goes to infinity as well. Specifically, because of the existence of the proportional delay factor r, the time evolution of the different Lur'e network is proportionally shifted in the unit rt instead of being linearly transformed by normal time delay described in most previous works. With that in mind, the methods used in [24, 26, 27] such as the impulsive comparison principle cannot be straightly used to this chapter. In this chapter, because of the proportional delay problem, the difficulty of calculating the upper bound of $v(t)$ in (2.21) is greatly raised. In most previous works discussing synchronization of complex networks with proportional delays or the stability of systems, such as [33–36], by proposing a kind of variable transformations $y_i(t) = z_i(e^t)$, the unbounded proportional time-delay system is transformed into a constant time-delay system with variable parameters. From the angle of variable transformation, we could discover that it would make the general time scale t become an exponential time scale e^t. Therefore, the systems in complex networks are exponentially shifted by the variable coefficient e^t instead of the normal time instant t, which largely increases the complexity and difficulty in calculation to a great extent. In order to solve the problem described before, in this chapter, by applying the impulsive comparison principle, the comparison system with proportional delay has been derived, and after that the method of variation of parameters and the average impulsive interval have been jointly used. Some necessary parameter equations like (2.24), $\bar{h}(t)$ and $H(t)$ are ingeniously constructed. Finally, the upper bound of $v(t)$ described in (2.21) is given precisely.

Remark 2.3 In fact, impulsive control is the most superior discontinuous control method, which provides instantaneous motions for the system. Compared with the conventional continuous control protocol, it could save the control cost greatly. In the research field of impulsive control, a very significant parameter is the impulsive effect μ. It could greatly affect the synchronization of complex networks. In the past, most of the researches on the stability of systems under impulsive control or the synchronization of complex networks only select some positive function impulsive effects which are beneficial to the final results in a limited range of values [25–31]. However, in order to broaden the application field of impulsive control, a wider range of impulsive effects should be considered [24, 32]. In this chapter, because the concept of average impulsive interval can be used, based on the functions in networks synchronization, the different impulsive value ranges of impulsive effects are studied. Especially, firstly, if the parameter ϖ that is closely related to the impulsive effects satisfies $0 < \varpi \leq 1$, then the impulses will have a positive effect on network synchronization. Considering the impact of this class of impulsive effects, we could take the feedback strength k_i as small as possible or k_i could be even taken as zero. Secondly, considering the other case, if the parameter $\varpi > 1$, then the impulses will have a negative effect on network synchronization. Based on this case, in order to eliminate the side effects of the negative impulses, we have to select the feedback strength k_i as big constants. In addition, the choice of impulsive effect has some special value. If impulsive effect $\mu = -1$, then one obtains $e(t_k^+) = 0$, which is impossible to realize in the field of impulsive pinning control. If impulsive effect $\mu = 0$ or $\mu = -2$, then one has $\delta = (1+\mu)^2 = 1$, which would be regarded as exceptional situations of Case I.

Remark 2.4 In this chapter, an effective impulsive pinning controller is skillfully introduced to study the cluster synchronization problems on nonidentically coupled Lur'e networks. From the controller (2.4), it is worth noting that among different clusters only those Lur'e systems which directly linked will be pinned by the controller (2.4). Especially, in order to pin the Lur'e systems in the ϕ_i-th cluster, the term $-k_i\Gamma e_i(t)$ is applied. At the same time, to eliminate the negative impact of direct connected Lur'e systems in different clusters, the summation term $c\sum_{j=1}^{N} g_{ij}\Gamma s_{\phi_j}(rt)$ is skillfully introduced.

Remark 2.5 In the past, the impulsive interval N_a is usually set as the maximum or the minimum impulsive interval such as [17, 25]. This method could increase the conservatism of the results because N_a satisfies

$$\min_k\{t_k - t_{k-1}\} \le N_a \le \max_k\{t_k - t_{k-1}\}$$

under normal conditions. After introducing the concept of average impulsive interval and the impulsive interval (t, T), it is obvious that by jointly using the positive number N_a, the freedom indicator N_0 and the impulsive interval (t, T), the times of impulses could be evaluated precisely. With that in mind, the introducing of the average impulsive interval could greatly decreases the conservatism of results.

Remark 2.6 For some special purposes of complex dynamic network in practical application, cluster synchronization is introduced to decompose the whole network into several subgroups, called clusters. However, in previous works of studying the synchronization of different types of complex network clusters, not only the whole coupling matrix is required to satisfy the diffusion condition, but also each block corresponding to the network cluster in the coupling matrix is required to satisfy the diffusion condition [12–15]. For example, [15] presented a matrix class $\mathbf{A}_1$ in Definition 2.3. That is for an $N \times N$ matrix $A = [A_{ij}]$ with blocks $A_{ii} \in \mathbb{R}^{(m_i - m_i - 1)\times(m_i - m_i - 1)}$ and $A_{ij} \in \mathbb{R}^{(m_i - m_i - 1)\times(m_j - m_j - 1)}$ $i, j = 1, \ldots, k$, A_{ij} satisfies the diffusive condition. Different from the previous works, the impulsive pinning controller is proposed in this chapter to eliminate the assumptions like zero-row-sum condition. It means that the theoretical results of this paper can be applied to more complex dynamic network types. The next section demonstrates this discussion with an example.

2.4 Numerical Simulations

Consider the Chua's circuits in differential equations form as below

$$\begin{cases} \dot{z}_1(t) = -a \cdot z_1(t) + a \cdot z_2(t) - a \cdot f(z_1(t)), \\ \dot{z}_2(t) = z_1(t) - d \cdot z_2(t) + d \cdot z_3(t), \\ \dot{z}_3(t) = -bz_2(t) - c \cdot z_3(t), \end{cases} \tag{2.30}$$

with $a = 9.78, b = 14.97, c = 0, d = 1$, and the nonlinear function is described as $f(z_1(t)) = 0.75z_1(t) + \frac{1}{2}(1.31 - 0.75)(|z_1 + 1| - |z_1 - 1|)$.

The Rössler oscillator

$$\begin{cases} \dot{s}_1(t) = -\alpha s_1(t) + \beta s_3(t), \\ \dot{s}_2(t) = \gamma z_1(t) + 0.2s_2(t), \\ \dot{s}_3(t) = 0.2 - 5.7s_3(t) + s_2(t) \cdot s_3(t) \end{cases} \tag{2.31}$$

with $\alpha = 1, \beta = 1, \gamma = 1$.

Firstly, we show the phase graphs of the two different Lur'e systems (2.30) and (2.31) in Fig. 2.1.

In this section, for better numerical simulation, the complex network designed by us consists of two clusters and six Lur'e systems. The first cluster consists of three Lur'e systems, which are modeled by the Chua's circuits. The second cluster consists of the left three systems, which are described by the Rössler oscillator. The Chua's circuits and the Rössler oscillator are shown in (2.30) and (2.31), respectively. It is worth noting that the third system (Chua's circuits) in the first cluster and the forth system (Rössler system) in the second cluster have a direct connection between the two clusters. Base on the above statement, by taking the structure of the network into consideration, one obtains $\phi_1 = \phi_2 = \phi_3 = 1, \phi_4 = \phi_5 = \phi_6 = 2, N = 6, l = 2$, and for clusters the set are described as $U_1 = \{1, 2, 3\}$, $U_2 = \{4, 5, 6\}$, $\bar{U}_1 = \{3\}$, $\bar{U}_2 = \{4\}$. The coupling matrix is designed as below

$$G = \begin{pmatrix} -1 & 0 & 1 & 0 & 0 & 0 \\ 0 & -1 & 1 & 0 & 0 & 0 \\ 1 & 1 & -3 & 1 & 0 & 0 \\ 0 & 0 & 1 & -3 & 1 & 1 \\ 0 & 0 & 0 & 1 & -1 & 0 \\ 0 & 0 & 0 & 1 & 0 & -1 \end{pmatrix},$$

where the coupling strength $c = 1$. Unlike most existed works, the delay considered in this paper is proportional delay. Let $r = 0.1$.

In the following, based on the different functions of the impulsive effects μ in cluster synchronization, three examples will be given. Considering the first cluster, the errors for each state are described as $E_1^j(t) = \sqrt{\frac{1}{3}((e_1^j(t))^2 + (e_2^j(t))^2 + (e_3^j(t))^2)}$

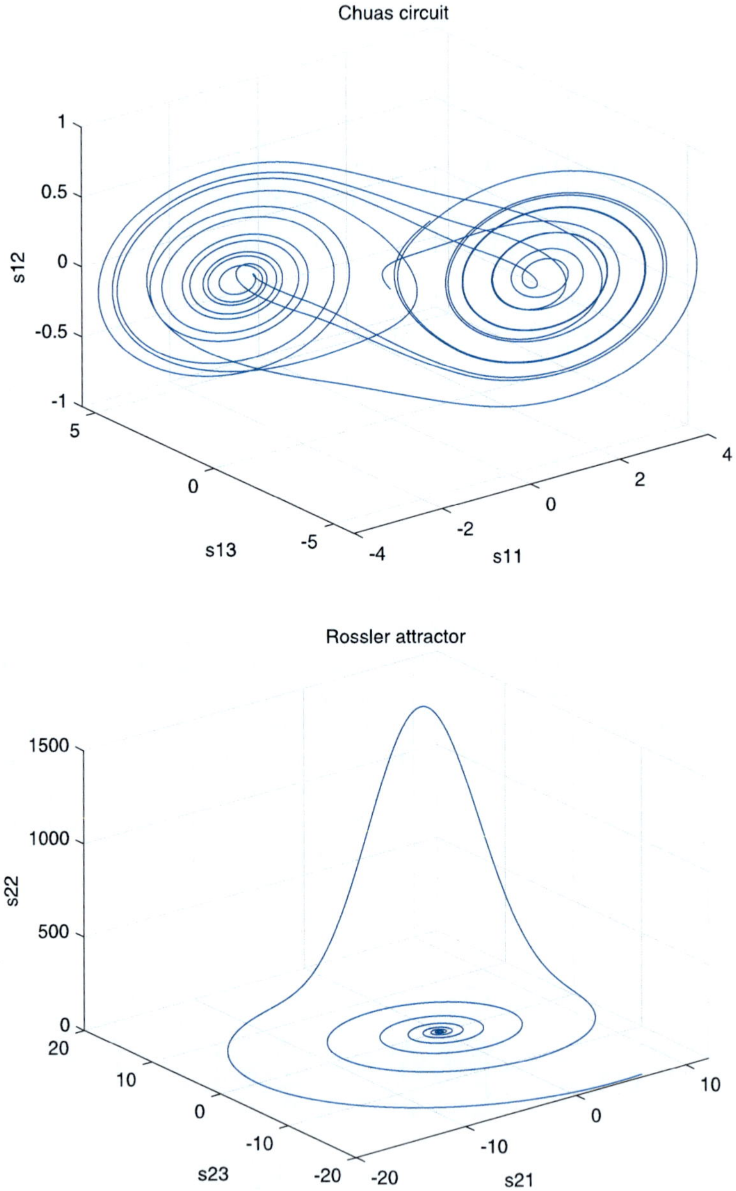

Fig. 2.1 The Chua's circuits and the Rössler system

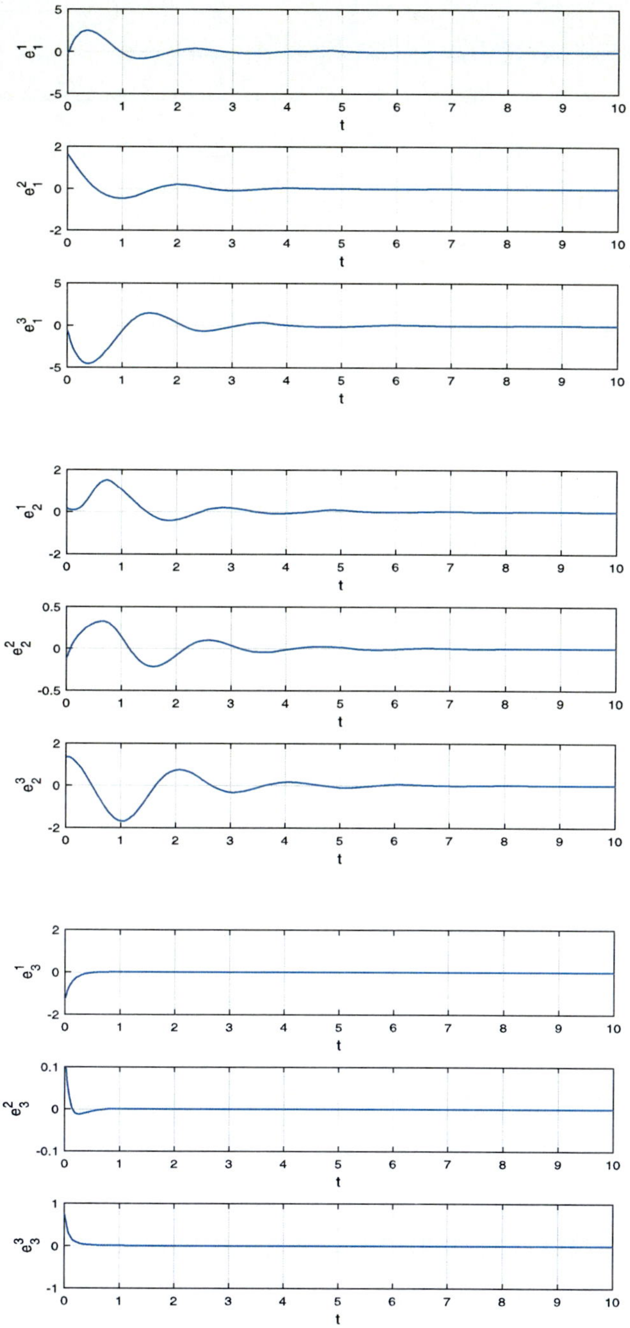

Fig. 2.2 The state curves of each error vector

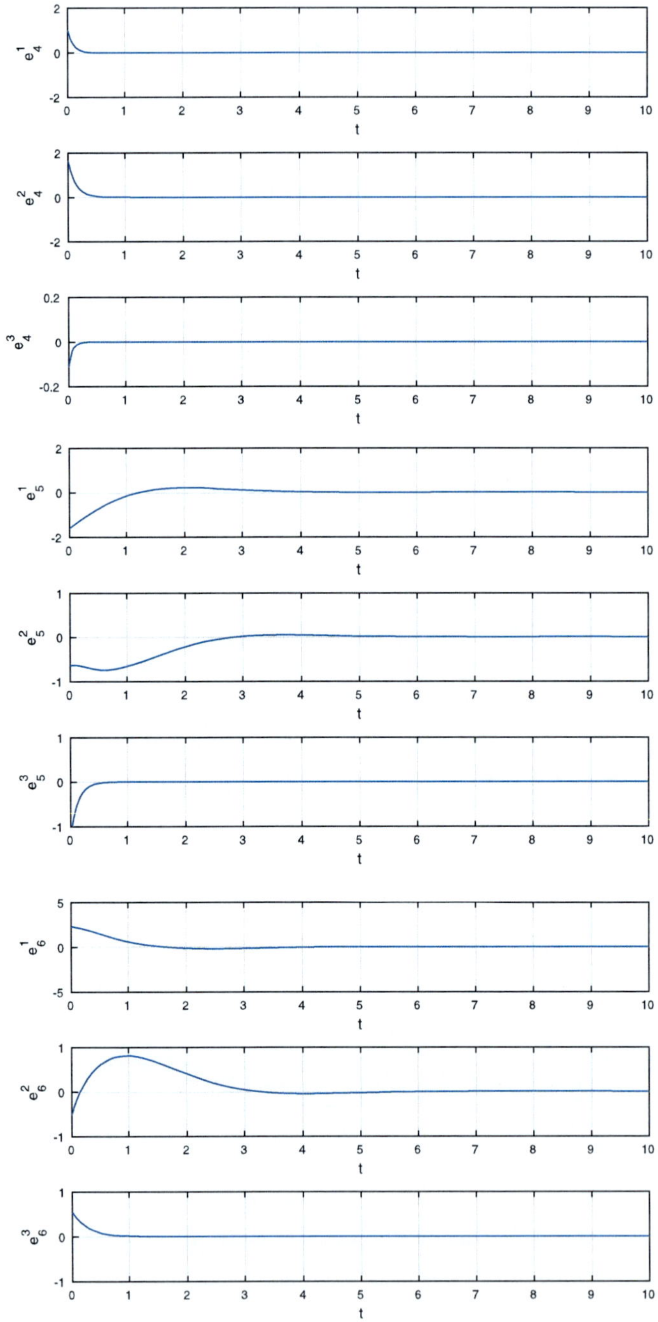

Fig. 2.3 The state curves of each error vector

and for the second cluster, the errors for each state are modelled as $E_2^j(t) = \sqrt{\frac{1}{3}((e_4^j(t))^2 + (e_5^j(t))^2 + (e_6^j(t))^2)}$ for $j = 1, 2, 3$.

Example 2.1 In the first example, set the impulsive effect $\mu = 0.2$. Then by simple calculation, we have $\varpi = (1 + \mu)^2 = 1.44$ satisfying $\varpi > 1$, which belongs to Case II in Theorem 2.1. Let the control gains be $k_1 = k_2 = k_5 = k_6 = 0$ and $k_3 = 10$, $k_4 = 8$. Based on the parameters introduced before, the conditions of Case II are satisfied. According to the description of $e_i(t) = z_i(t) - s_{\phi_i}(t)$ for $i = 1, 2, \ldots, 6$, we could have the six error vectors for the six Lur'e systems in the complex networks. Figures 2.2 and 2.3 give the curves of the six error vectors, where we can easily conclude that each state of the error vector approaches to zero. In addition, Fig. 2.4 shows the error curves of the network corresponding to two different clusters. From the above figures, we could easily conclude that by introducing the impulsive pinning control protocol (2.4), the objective of cluster synchronization of the nonidentically coupled Lur'e networks with proportional delay rt and impulsive effect $\mu = 0.2$ is achieved in the end.

Example 2.2 In this example, set the impulsive effect factor be $\mu = -0.2$, and we have $\varpi = 0.64 \in (0, 1)$, which satisfies Case I in Theorem 2.1. With similar procedure in Example 2.1, in Fig. 2.5, we give the error state graphs of two different clusters, which means the cluster synchronization on the nonidentically coupled Lur'e networks with proportional delay rt and impulsive effect factor $\mu = -0.2$ under the impulsive pinning controller (2.4).

Example 2.3 In this example, we will illustrate the efficiency of the feedback term $k_i \Gamma e_i(t)$ that is a parameter of the impulsive pinning controller (2.4) in detail, the impulsive effect μ is modelled as -0.2, and the feedback control strength is described as $k_i = 0$ $(i = 1, 2, \ldots, 6)$. We show the curves of different cluster errors in Fig. 2.6 by the parameters introduced before. Then we can find that all error states can not go to zero except the third error state in the second cluster. The result means with a short time period, the nonidentically coupled nonlinear Lur'e networks cannot achieve cluster synchronization, and this is further illustrated that because of the impact of the negative feedback term $-k_i \Gamma e_i(t)$, the negative effects cased by the impulsive effects would be removed.

Remark 2.7 In this simulation, we set the coupling matrix G that satisfies the diffusive condition. Denote

$$G = \begin{pmatrix} -1 & 0 & 1 & 0 & 0 & 0 \\ 0 & -1 & 1 & 0 & 0 & 0 \\ 1 & 1 & -3 & 1 & 0 & 0 \\ 0 & 0 & 1 & -3 & 1 & 1 \\ 0 & 0 & 0 & 1 & -1 & 0 \\ 0 & 0 & 0 & 1 & 0 & -1 \end{pmatrix} = \begin{pmatrix} \bar{G}_{11} & \bar{G}_{12} \\ \bar{G}_{21} & \bar{G}_{22} \end{pmatrix}$$

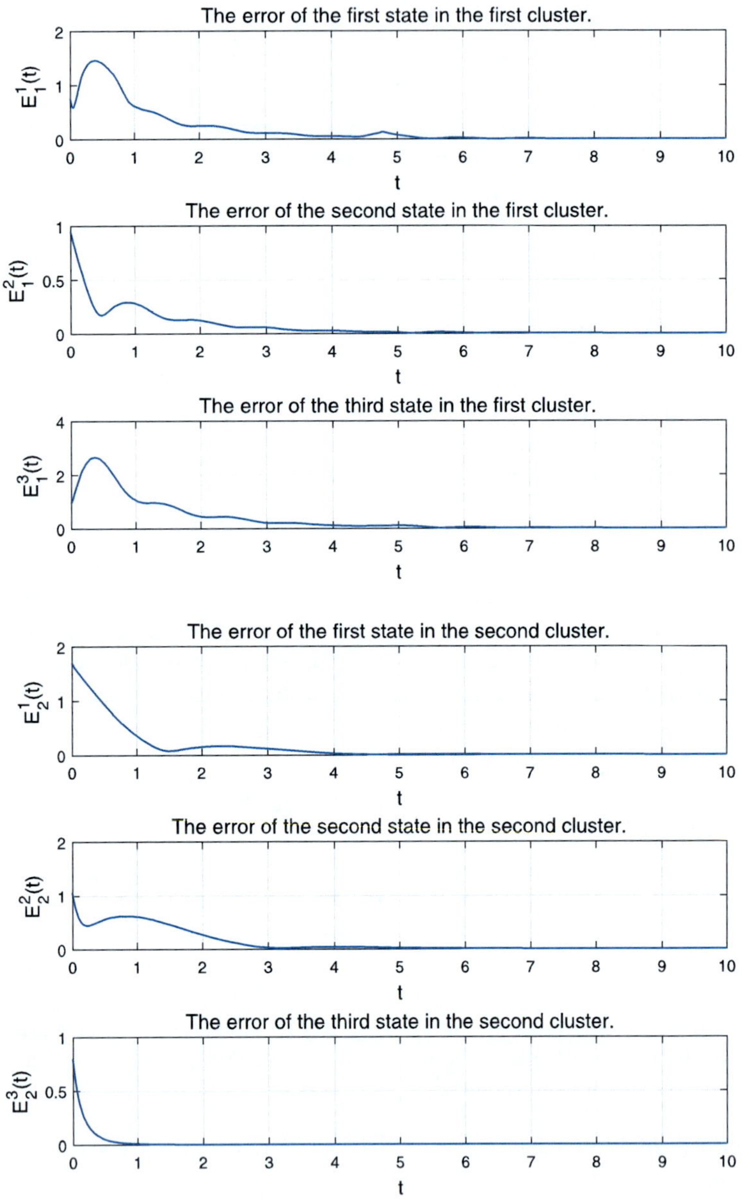

Fig. 2.4 The evolution curves of the error in two clusters with $\mu = 0.2$

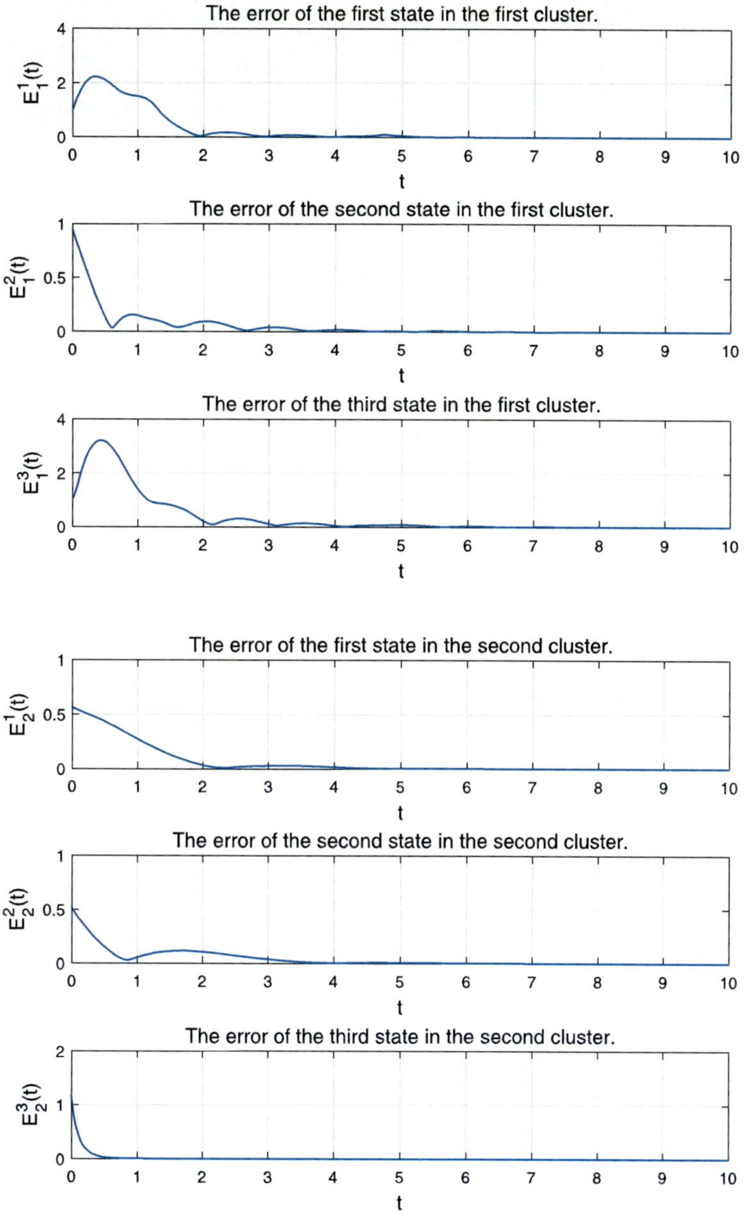

Fig. 2.5 The evolution curves of the error in two clusters with $\mu = -0.2$

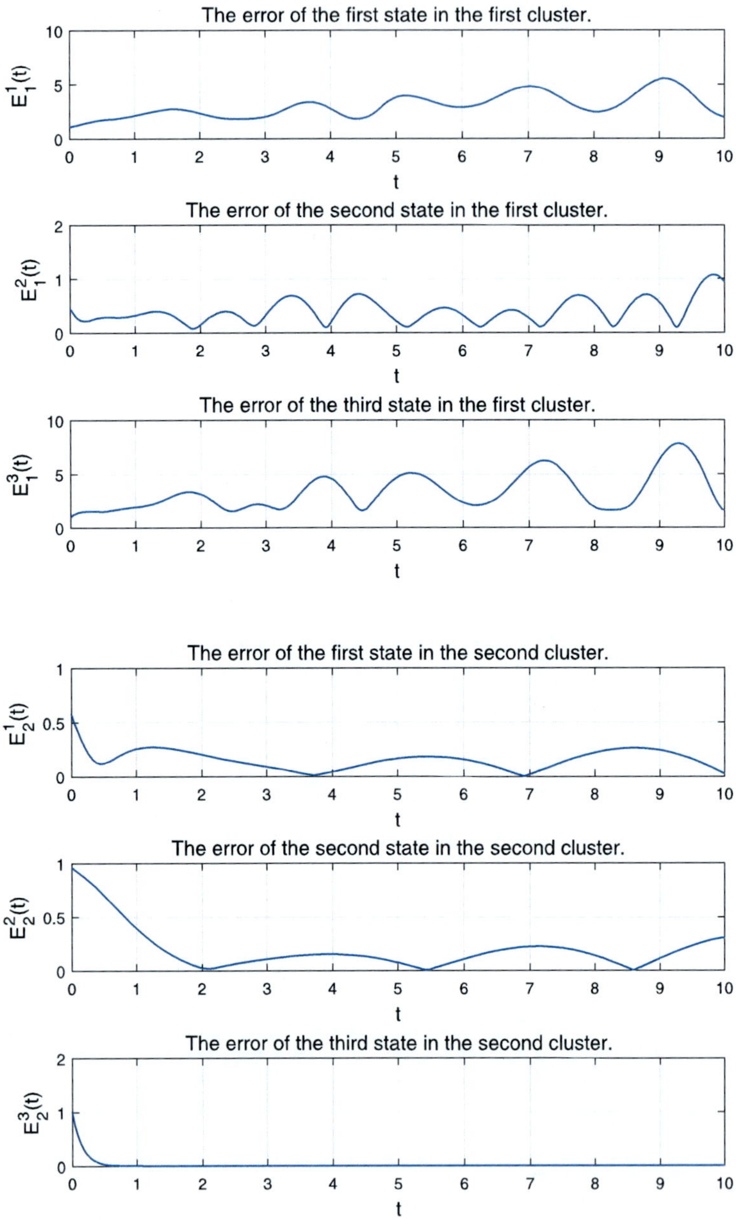

Fig. 2.6 The evolution curves of the error in two clusters with $\mu = -0.2$ and $k_i = 0(i = 1, \ldots, 6)$

where the block matrix $G_{ij} \in \mathbb{R}^{3\times 3}$ for $i, j = 1, 2$. G_{11} and G_{22} correspond to two different clusters in the Lur'e networks, respectively. It is worth noting that G_{11} and G_{22} do not necessarily satisfy the zero-row-sum condition. In other words, we could conclude that each block matrix of the outer-coupling matrix does not necessarily satisfy the diffusive condition.

2.5 Conclusion

In this chapter, by designing the effective impulsive pinning controller, cluster synchronization of a kind of nonidentically coupled Lur'e dynamical networks with proportional delay has been studied. Because of the characteristic of unbounded in the proportional delay, the difficulty and complexity for cluster synchronization have been increased to a great extent. Considering the cluster tree topology of Lur'e network, the impulsive pinning controller is applied to the Lur'e system directly connected between different clusters. Some sufficient conditions for the cluster synchronization of nonidentically coupled Lur'e networks have been derived by introducing the method of variation of parameters, the impulsive comparison principle with proportional delay and the definition of the average impulsive interval and considering different impacts of the impulsive effects on cluster synchronization. On this basis, the exponential convergence rate is estimated successfully. Finally, three numerical simulations have strongly proved the rationality of the proposed theory.

References

1. Meyerhenke, H., Sanders, P., Schulz, C.: Parallel graph partitioning for complex networks. IEEE Transactions on Parallel and Distributed Systems. **28**, 2625–2638 (2017)
2. Castellano, C., P-Satorras, R.: Relating topological determinants of complex networks to their spectral properties: structural and dynamical effects. Physical Review X. **7**, 041024:1–12 (2017)
3. Lu, R.Q., Yu, W.W., Lv, J.H.: Synchronization on complex networks of networks. IEEE Transactions on Neural Networks and Learning Systems. **25**, 2110–2118 (2014)
4. Yang, X.S., Lam, J., Ho, D.W.C.: Fixed-time synchronization of complex networks with impulsive effects via nonchattering control. IEEE Transactions on Automatic Control. **62**, 5511–5521 (2017)
5. Wang, Z.H., Xu, J.J., Song X.M.: Consensus problem in multi-agent systems under delayed information. Neurocomputing. **316**, 277–283 (2018)
6. Wen, G.H., Wan, Y., Cao, J.D.: Master slave synchronization of heterogeneous systems under scheduling communication. IEEE Transactions on Systems, Man, and Cybernetics: Systems. **48**, 473–484 (2018)
7. Song, X.M., Park, J.H.: Linear optimal estimation for discrete-time measurement-delay systems with multi-channel multiplicative noise. IEEE Transactions on Circuits and Systems II-Express Briefs. **64**, 156–160 (2017)
8. Huo, X. Ma, L., Zhao, X.D.: Observer-based fuzzy adaptive stabilization of uncertain switched stochastic nonlinear systems with input quantization. Journal of the Franklin Institute. **356**, 1789–1809 (2019)

9. He, W.L., Chen, G.R., Han, Q.L.: Multiagent systems on multilayer networks: synchronization analysis and network design. IEEE Transactions on Systems, Man, and Cybernetics: Systems. **47**, 1655–1667 (2017)
10. Zhang, Z.M., He, Y., Wu, M.: Exponential synchronization of neural networks with time-varying delays via dynamic intermittent output feedback control. IEEE Transactions on Systems, Man, and Cybernetics: Systems. **49**, 612–622 (2019)
11. Zhang, D., Wang, Q.G., Srinivasan, D.: Asynchronous state estimation for discrete-time switched complex networks with communication constraints. IEEE Transactions on Neural Networks and Learning Systems. **29**, 1732–1746 (2018)
12. Zhou, L.L., Wang, C.H., Du, S.C.: Cluster synchronization on multiple nonlinearly coupled dynamical subnetworks of complex networks with nonidentical nodes. IEEE Transactions on Neural Networks and Learning Systems. **28**, 570–583 (2017)
13. Su, H.S., Rong, Z.H., Chen, M.Z.Q.: Decentralized adaptive pinning control for cluster synchronization of complex dynamical networks. IEEE Transactions on Cybernetics. **43**, 394–399 (2013)
14. Yang, X.S., Ho, D.W.C., Lu, J.Q.: Finite-time cluster synchronization of T-S fuzzy complex networks with discontinuous subsystems and random coupling delays. IEEE Transactions on Fuzzy Systems. **23**, 2302–2316 (2015)
15. Li, L.L., Ho, D.W.C., Cao, J.D.: Pinning cluster synchronization in an array of coupled neural networks under event-based mechanism. Neural Networks. **76**, 1–12 (2016)
16. Liu, X.W., Chen, T.P.: Cluster synchronization in directed networks via intermittent pinning control. IEEE Transactions on Neural Networks. **22**, 1009–1020 (2011)
17. Lu, J.D., Wang, Z.D., Cao, J.D.: Pinning impulsive stabilization of nonlinear dynamical networks with time-varying delay. International Journal of Bifurcation and Chaos. **22**, 1250176:1–12 (2012)
18. Yang, Z.C., Xu, D.Y.: Stability analysis and design of impulsive control systems with time delay, IEEE Transactions on Automatic Control. **52**, 1448–1454 (2007)
19. Chen, W.H., Lu, X., Zheng, W.X.: Impulsive stabilization and impulsive synchronization of discrete-time delayed neural networks. IEEE Transactions on Neural Networks and Learning Systems. **26**, 734–748 (2015)
20. Lakshmikantham, V., Bainov, D., Simenonv, P.: Theory of impulsive different Equations. World Scientific. Singapore (1989)
21. Yang, Z.C., Xu, D.Y.: Stability analysis of delay neural networks with impulsive effects. IEEE Transactions on Circuits and Systems II: Express Briefs. **52**, 517–521 (2005)
22. Lu, J.D., Ho, D.W.C., Cao, J.D.: Exponential synchronization of linearly coupled neural networks with impulsive disturbances. IEEE Transactions on Neural Networks. **22**, 329–336 (2011)
23. Tang, Z., Park, J.H., Lee, T.H.: Mean square exponential synchronization for impulsive coupled neural networks with time-varying delays and stochastic disturbances. Complexity. **21**, 190–202 (2016)
24. Tang, Z., Park, J.H., Feng, J.W.: Impulsive effects on quasi-synchronization of neural networks with parameter mismatches and time-varying delay. IEEE Transactions on Neural Networks and Learning Systems. **29**, 908–919 (2018)
25. Hu, J.Q., Liang, J.L., Cao, J.D.: Synchronization of hybrid-coupled heterogeneous networks: pinning control and impulsive control schemes. Journal of the Franklin Institute. **351**, 2600–2622 (2014)
26. He, W.L., Qian, F., Han, Q.L.: Lag quasi-synchronization of coupled delayed systems with parameter mismatch. IEEE Transactions on Circuits and Systems I: Regular Papers. **58**, 1345–1357 (2011)
27. He, W.L., Qian, F., Lam, J.: Quasi-synchronization of heterogeneous dynamic networks via distributed impulsive control: Error estimation, optimization and design. Automatica. **62**, 249–262 (2015)
28. He, W.L., Chen, G.R., Han, Q.L.: Network-based leader-following consensus of nonlinear multi-agent systems via distributed impulsive control. Information Sciences. **380**, 145–158 (2017)

29. Zhang, W.B., Tang, Y., Miao, Q.Y.: Synchronization of stochastic dynamical networks under impulsive control with time delays. IEEE Transactions on Neural Networks and Learning Systems. **25**, 1758–1768, (2014)
30. Wong, W.K., Zhang, W.B., Tang, Y.: Stochastic synchronization of complex networks with mixed impulses. IEEE Transactions on Circuits and Systems I: Regular Papers. **60**, 2657–2667 (2013)
31. Ma, T.D., Lewis, F.L., Song, Y.D.: Exponential synchronization of nonlinear multi-agent systems with time delays and impulsive disturbances. International Journal of Robust and Nonlinear Control. **26**, 1615–1631 (2016)
32. Lu, J.Q., Ho, D.W.C., Cao, J.D.: A unified synchronization criterion for impulsive dynamical networks. Automatica. **46**, 1215–1221 (2010)
33. Zhou, L.Q., Zhang, Y.Y.: Global exponential stability of cellular neural networks with multi-proportional delays. International Journal of Biomathematics. **8**, 1550071:1–17 (2015)
34. Jia, R.W.: Finite-time stability of a class of fuzzy cellular neural networks with multi-proportional delays. Fuzzy Sets and Systems. **319**, 70–80 (2017)
35. Zhou, L.Q.: Delay-dependent exponential synchronization of recurrent neural networks with multiple proportional delays. Neural Processing Letters. **42**, 619–632 (2015)
36. Zhou, L.Q., Chen, X.B., Yang, Y.X.: Asymptotic stability of cellular neural networks with multiple proportional delays. Applied Mathematics and Computation. **229**, 457–466 (2014)

Chapter 3
Impulsive Synchronization of Derivative CNNs with Cluster-Tree Topology

3.1 Introduction

In the past two or three decades, the research on complex systems and complex dynamical networks (CDNs) in the field of information and science has risen to a high degree of enthusiasm [1–5]. Many research branches have been proposed for CDNs, for instance, the transmission mechanisms and dynamics, the community structures, the searching mechanisms, the topology structures and characters and the synchronization of the CDNs. There is no doubt that synchronization of complex networks is one of the most popular directions in these topics on complex dynamical networks. In fact, the synchronization describes a phenomenon, where the motion trajectories of some or all systems in a complex network move closer to the desired system or goal trajectory. The synchronization of CDNs has been applied to various fields, for instance, power systems, biopharmaceuticals and financial network, social network and so on, see [6–16] and references therein.

In reality, except for a few complex networks that can be synchronized by adjusting their system parameters, e.g., coupling weights and coupling strength [6], most complex networks are not possible when there is no external input. Accordingly, designing the suitable controller is essential for the synchronization of complex networks [8, 9]. After decades of development, many excellent control strategies have been proposed [10]. These control protocols could be roughly divided into two categories from the perspective of time series, discrete-time control methods, impulsive control, intermittent control and the continuous-time control protocols, for instance the pinning feedback control, adaptive control. Actually, the impulsive control is a kind of superior discontinuous control strategy which only actively works at each impulsive instant in a very sparse sequence of time. From this perspective, impulsive control greatly decrease control costs and be applied to investigate synchronization well [17–26]. Through the efforts of the past few decades, the research on single impulsive control protocol has been very mature and the joint use with other control methods has become the research focus. For example, the pinning impulsive methods, the stabilization issues of a class of nonlinearly time delay coupled networks in

Z. Tang et al., *Impulsive Synchronization of Complex Dynamical Networks*,
https://doi.org/10.1007/978-981-16-5383-4_3

view of the Lyapunov method and the comparison principle has been studied by Lu et al. in [17]. Additionally, Zhang et al. reached the stochastic synchronization for the CDNs with delays and by designing delayed impulsive controller, several sufficient conditions were acquired which ensure the exponential stable of the error CDNs in [19].

In some practical projects, sometimes it is only necessary to synchronize the systems in each sub-network instead of the entire network. For this situation that complete synchronization for the complex networks would cause the waste of resources. In order to solve this problem, it is very important to introduce cluster synchronization. It only need to synchronize in the same cluster when there is no requirement between any two different clusters [27–31]. During the past years, complex networks or multi-agents with cluster-tree topologies have also been focused. Such as the finite time synchronization of nonlinear coupled complex networks with cluster-tree topology and to has been investigated by Yang et al. in [28]. Moreover, in our previous work [30], we have studied the cluster synchronization of complex networks, which is obtained in a finite time by using some mathematical methods cleverly.

Additionally, In order to reflect the real situations of CDNs not only in practical applications but also in artificial society, it is not enough to simulate the information exchange among systems by only considering the coupling at time states or/and time delay states. In some special scenes, couplings network model at time derivative states should be considered [32–34]. For the coupled neural networks (CNNs) with state vector couplings and derivative couplings, it can be shown that the rate of the i-th state is used to describe the spatial position of the i-th neural network (NN), while the derivative on state vectors could be deemed as the rate of information exchange among different neural networks (NNs). Therefore, from the description of the derivative coupled networks, it could be obtained that the rate of the i-th NN is decided by the dynamics of itself and the rates of its neighbor NNs. Zheng et al. in [33] has investigated the problem of synchronization for a class of complex networks with and without derivative couplings under the pinning impulsive controller. In addition, synchronization for the complex networks without derivative and with derivative couplings have been addressed in [34], where pinning synchronization conditions were obtained.

As far as we know, the cluster synchronization problem of derivative CNNs with time-varying delays and stochastic disturbance, until now, receives few concerns. The complexity in theoretical proof and the importance in practical engineering applications motivate us to do this work. In this chapter, we study the exponential cluster synchronization for a kind of derivative coupling NNs with time-varying delay. Moreover, because the network is always disturbed by some uncertain factors, to simulate the actual situation, a stochastic disturbance is discussed when modeling the networks. Distinct characteristics of this chapter can be concluded as follow: (1) Three different coupling methods, namely, general state coupling, delayed state coupling and state derivative coupling are simultaneously considered when modeling the coupled NNs; (2) The CNNs with cluster-tree topology structure is discussed and an impulsive pinning controller is designed, which only imposed on the NNs in the cluster that have directed connections with the NNs in other clusters; (3) Due to the existence of derivative couplings, a new Lyapunov function is constructed, which closely related to the derivative coupling matrix and coupling strength of the derivative. In addition, general comparison principle and mathematical induction are skillfully extended to smooth the theoretical proof; (4) By introducing the concept of average impulsive interval, the functions with different impulsive effects are analyzed uniformly. Simultaneously, different from general time delays, impulse delay is also considered while designing the impulsive pinning controller.

The rest of paper is arranged as follows. In Sect. 3.2, the derivative CNNs model is presented and some definitions, lemmas and assumptions are provided. Then the mean square synchronization of the derivative CNNs is studied in Sect. 3.3. In the Sect. 3.4, in order to illustrate the correctness of the main theoretical analysis results, a numerical simulation is proposed. In Sect. 3.5, we will draw the conclusion of this chapter.

Notations. G^T is expressed as the transpose of the matrix G. $\mathbb{R}^n$ stands for the n-dimensional Euclidean space. $\mathbb{R}^{n\times n}$ is the set of $n \times n$ real matrices. The symbol $\|\cdot\|$ denotes the Euclid norm of the matrix or the vector. diag$\{\cdots\}$ denotes a diagonal matrix. $\lambda_{\max}(G)$ denotes the largest eigenvalue of matrix G, and $\max_{1\le i\le M}\{\cdot\}$ expresses to take the maximum value. $B = [b_{ij}] \in \mathbb{R}^{m\times m}$ stands for the matrix with elements b_{ij}. We denote the $G > 0 (G \geq 0)$, which is positive definite (semi-definite). I_m represents the m-dimensional identity matrix $G = (g_{ij})$ denotes a $\mathbb{R}^{M\times M}$ matrix G with elements g_{ij} for $i, j = 1, 2, \ldots, M$. $Trace(G)$ denotes the trace of matrix H. The set $\mathfrak{N}^+$ denotes the positive integers. $D^+v(t) = \overline{\lim}_{t->0^+} \frac{v(t+\sigma)-v(t)}{\sigma}$. It assumes that the matrices meet the requirements of algebraic operations in this chapter.

3.2 Model Description and Preliminaries

3.2.1 Network Structure Statement

Before starting the work, we give the following assumptions for derivative CNNs with cluster-tree topology structure. Consider the coupled NNs with total M NN and the number of the clusters q satisfying $M > q \geq 2$. If the i-th NN belongs to the j-th cluster, it will be denoted as $\phi_i = j$. Let the set Ω_j be all NNs in the j-th cluster and the set $\bar{\Omega}_j$ be all NNs in the j-th cluster which directly connected to the NN in other clusters. Through the above analysis, we can get (i) $\Omega_i \bigcap \Omega_j = \emptyset$, for $i \neq j$ and $i, j = 1, 2, \ldots, q$; (ii) $\bigcup_{i=1}^{l} \Omega_i = \{1, 2, \ldots, M\}$.

3.2.2 Problem Formulation

Consider the controlled derivative CNNs with q clusters and stochastic disturbance

$$
\begin{aligned}
dz_i(t) = [&A_{\phi_i} z_i(t) + B_{\phi_i} f^1_{\phi_i}(z_i(t)) + C_{\phi_i} f^2_{\phi_i}(z_i(t-\tau(t))) + \tilde{I}(t) + c_1 \sum_{j=1}^{N} l_{ij}\Gamma z_j(t) \\
&+ c_2 \sum_{j=1}^{N} q_{ij}\Gamma z_j(t-\bar{\tau}(t)) + c_3 \sum_{j=1}^{N} g_{ij}\Gamma \dot{z}_j(t) + u_i(t)]dt \\
&+ \tilde{h}(t, z_i(t), z_i(t-\tau(t)), z_i(t-\bar{\tau}(t))d\tilde{\omega}(t)
\end{aligned} \tag{3.1}
$$

where $z_i(t) = [z_i^1(t), z_i^2(t), \ldots, z_i^m(t)]^T \in \mathbb{R}^m$ is the i-th NN state vector. Matrix $A_{\phi_i} = \text{diag}\{a_{\phi_i 1}, \ldots, a_{\phi_i m}\} \in \mathbb{R}^{m\times m}$ is a positive definite matrix, where $a_{\phi_i j}$ represents the velocity of the j-th neural cell resets its potential to the resting sate when isolated from other neural cells and external inputs. $B_{\phi_i} = [b_{\phi_i dj}] \in \mathbb{R}^{m\times m}$ and $C_{\phi_i} = [c_{\phi_i dj}] \in \mathbb{R}^{m\times m}$ are constant matrices, where $b_{\phi_i dj}$ and $c_{\phi_i dj}$ denote the strengths of connectivity at time t and $t-\tau(t)$, respectively. $f^d_{\phi_i} : \mathbb{R}^m \to \mathbb{R}^m$ are called the activation functions, which are continuously differentiable on $\mathbb{R}$ for $d = 1, 2$. $\tilde{I}(t) \in \mathbb{R}^m$ is an external input vector; $\tau(t)$ and $\bar{\tau}(t)$ are system time-varying delay and communication time-varying delay, respectively. The positive parameters c_1, c_2, c_3 are the coupling strengths and matrix $\Gamma = \text{diag}\{r_1, \ldots, r_n\} \in \mathbb{R}^{m\times m}$ is the inner connection matrix satisfying $r_i \geq 0$. Additionally, it assumes that $\Gamma = I_m$ for simplicity in this chapter. $L = [l_{ij}] \in \mathbb{R}^{M\times M}$ and $Q = [q_{ij}] \in \mathbb{R}^{M\times M}$ are decided by Lur'e network topology structures and considered to be asymmetric and irreducible in this chapter. Additionally, $l_{ij}(\neq l_{ji}) > 0$ $(q_{ij}(\neq q_{ji}) > 0)$ if and only if there is a connection between from the j-th NN to the i-th NN for $i \neq j$, and $l_{ij} = 0$ $(q_{ij} = 0)$ otherwise. As well as matrices B, W meet zero-sum-row conditions, that is, $l_{ii} = -\sum_{j\neq ij=1}^{M} l_{ij}$ and $q_{ii} = -\sum_{j\neq ij=1}^{M} q_{ij}$. $G = [g_{ij}] \in \mathbb{R}^{M\times M}$ is the derivative coupled matrix and satisfies the diffusive condition, namely, $g_{ij} = g_{ji} > 0$ if there are connections between the j-th NN and the i-th NN with $i \neq j$, and $g_{ij} = g_{ji} = 0$ otherwise. $\tilde{\omega}(t) \in \mathbb{R}^n$ is an n-dimensional Brownian motion. $\tilde{h} : \mathbb{R}^+ \times \mathbb{R}^m \times \mathbb{R}^m \times \mathbb{R}^m \to \mathbb{R}^{m\times n}$ is called the noise intensity matrix-valued function which assumed to be $\tilde{p}(t, 0, 0, 0) = \mathbf{0}$. $u_i(t)$ is the controller, which will be designed later.

Define $\mathcal{C}([-\tilde{\tau}, 0], \mathbb{R}^m)$ as the Banach space from $[-\tilde{\tau}, 0]$ to $\mathbb{R}^m$, where $\tilde{\tau} = \max\{\tau(t), \bar{\tau}(t)\}$. For the derivative CNNs (3.1), its initial values are provided by $z_i(t) = \omega_i(t) \in \mathcal{C}([-\tilde{\tau}, 0], \mathbb{R}^m)$.

Remark 3.1 From the definition of the derivative coupled matrix G, the left eigenvalue of matrix G can be denoted and arranged as $0 = \lambda_1 \geq \lambda_2 \geq \cdots \geq \lambda_M$. It implies that the matrix $I_M - c_3 G$ is a positive definite matrix with the maximum eigenvalues $\lambda_{\max}(I_M - c_3 G) \geq 1$ and $\lambda_{\min}(I_M - c_3 G) = 1$. This fact plays an indispensable role in the construction of the Lyapunov function later, and also affects the classification of the impulsive effects.

In order to realize the global cluster synchronization, let the NNs with stochastic disturbance be the synchronization targets in each cluster

$$\begin{aligned} ds_{\phi_i}(t) = {} & [A_{\phi_i}s_{\phi_i}(t) + B_{\phi_i}f^1_{\phi_i}(s_{\phi_i}(t)) + C_{\phi_i}f^2_{\phi_i}(s_{\phi_i}(t-\tau(t))) + I(t)]dt \\ & + \tilde{h}(t, s_{\phi_i}(t), s_{\phi_i}(t-\tau(t)), s_{\phi_i}(t-\bar{\tau}(t))d\tilde{\omega}(t), \end{aligned} \tag{3.2}$$

where $s_{\phi_i}(t) = [s^1_{\phi_i}(t), s^2_{\phi_i}(t), \ldots, s^m_{\phi_i}(t)]^T \in \mathbb{R}^m$, $i = 1, 2, \ldots, M$. Moreover, the solution vector $s_{\phi_i}(t)$ of the NNs (3.2) could be deemed as the leader in the ϕ_i-th cluster and correspondingly, all NNs in the ϕ_i-th cluster could be considered as its followers. Namely, the cluster synchronization issue between the derivative CNNs (3.1) and the NNs (3.2) can be regarded as a special class of leader-following issues in each cluster.

Define the error vector as $\delta z_i(t) = z_i(t) - s_{\phi_i}(t)$ with $\delta z_i(t) = [\delta z^1_i(t), \delta z^2_i(t), \ldots, \delta z^m_i(t)]^T \in \mathbb{R}^m$ for $i = 1, 2, \ldots, M$. And design the impulsive pinning controller as

$$u_i(t) = u_{0,i}(t) + u_{1,i}(t) \tag{3.3}$$

where impulsive control part is designed as

$$u_{0,i}(t) = \sum_{k=1}^{+\infty}(\phi\delta z_i(t) + \mu\delta z_i(t - \check{\tau}(t)))\theta(t - t_k),$$

and for the pinning feedback control part $u_{1,i}(t)$, if $i \in \bar{\Omega}_{\phi_i}$

$$u_{1,i}(t) = -p_i\delta z_i(t) - c_1\sum_{j=1}^{N} l_{ij}s_{\phi_j}(t) - c_3\sum_{j=1}^{N} g_{ij}\dot{s}_{\phi_j}(t) - c_2\sum_{j=1}^{N} q_{ij}s_{\phi_j}(t - \bar{\tau}(t));$$

Otherwise, $u_{1,i}(t) = 0$. Nonnegative parameter p_i $(i = 1, 2, \ldots, M)$ is the control strength that can be suitably selected by the NNs (3.1); Let the control strength matrix be $P = \mathrm{diag}\{p_1, p_2, \ldots, p_m\}$; The parameters ϕ and μ represent the impulsive effects relate to the error state and delayed error state, respectively; $\theta(\cdot)$ is the Dirac impulsive function; For the impulsive signal, assume that it is a the strictly increasing sequence of time series $\xi = \{t_1, t_2, \ldots\}$ satisfying $t_{k-1} < t_k$ and $\lim_{k\to\infty} t_k = +\infty$ for $k \in \mathfrak{N}^+$ at the impulsive instants.

Through the discussion above, the controlled error NNs with derivative coupling and stochastic disturbance could be expressed as

$$\begin{cases} \dot{\delta z_i}(t) = [A_{\phi_i}\delta z_i(t) + B_{\phi_i}\Phi_{\phi_i}(\delta z_i(t)) + c_1\sum_{j=1}^{N} l_{ij}\delta z_j(t) \\ \quad +C_{\phi_i}\Psi_{\phi_i}(\delta z_i(t-\tau(t))) + c_2\sum_{j=1}^{N} q_{ij}\delta z_j(t-\bar{\tau}(t)) \\ \quad +c_3\sum_{j=1}^{N} g_{ij}\dot{\delta z_j}(t) - p_i\delta z_i(t)]dt, \\ \quad +h(t, \delta z_i(t)\delta z_i(t-\tau(t)), \delta z_i(t-\bar{\tau}(t))d\tilde{\omega}(t), \ t \neq t_k, \\ \Delta\delta z_i(t_k) = \phi\delta z_i(t_k^-) + \mu\delta z_i(t_k^- - \check{\tau}(t_k^-)), \\ \delta z_i(t) = \Lambda_i(t), \ -\tau \leq t \leq 0, \end{cases} \tag{3.4}$$

where $\tau = \max\{\tau(t), \bar{\tau}(t), \check{\tau}(t)\}$, functions $\Phi_{\phi_i}(\delta z_i(t)) = f_{\phi_i}^1(z_i(t)) - f_{\phi_i}^1(s_{\phi_i}(t))$, $\Psi_{\phi_i}(\delta z_i(t)) = f_{\phi_i}^2(z_i(t)) - f_{\phi_i}^2(s_{\phi_i}(t))$, $h(t, \delta z_i(t), \delta z_i(t-\tau(t)), \delta z_i(t-\bar{\tau}(t)) = \tilde{h}(t, z_i(t), z_i(t-\tau(t)), z_i(t-\bar{\tau}(t)) - \tilde{h}(t, s_{\phi_i}(t), s_{\phi_i}(t-\tau(t)), s_{\phi_i}(t-\bar{\tau}(t)), \quad i = 1, 2, \ldots, N$. $\Lambda_i(t)$ is the initial value of the error NNs for $t \in [-\tau, 0]$.

In all through the chapter, $\delta z_i(t)$ is supposed to be right-hand continuous at $t = t_k$, $k \in \mathfrak{N}^+$, and $\delta z_i(t_k) = \delta z_i(t_k^+) = \lim_{\epsilon \to 0^+} \delta z_i(t_k + \epsilon)$, $\delta z_i(t_k^-) = \lim_{\epsilon \to 0^-} \delta z_i(t_k + \epsilon)$. Then the solutions of the controlled error NNs (3.4) are piecewise right-hand continuous functions with discontinuities at $t = t_k$ for $k \in \mathfrak{N}^+$.

3.2.3 Preliminaries

To achieve the cluster synchronization on the derivative CNNs, we provide some definitions, lemmas and assumptions are necessary, which are necessary all throughout this chapter.

Definition 3.1 The derivative CNNs (3.1) is said be globally and exponentially stabilized to the leader NNs (3.2) if for any initial state $\psi_i(\cdot)(i = 1, 2, \ldots, M)$, there exist parameters $\lambda > 0$, $M_0 > 0$, $t_0 > 0$, such that

$$\mathbb{E}[\|z_i(t) - s(t)\|^2] \leq M_0 \sup_{s \in [t-\tau, t]} \mathbb{E}[\|\psi_i(s)\|^2] e^{-\lambda t} \tag{3.5}$$

for all $t > t_0 > 0$, Additionally, λ is called the exponential convergence velocity.

Definition 3.2 ([35]) For all initial conditions, if there holds $\lim_{t \to +\infty} \|z_i(t) - z_j(t)\| = 0 (i, j = 1, 2, \ldots, M)$ if and only if $\phi_i = \phi_j$ and $\lim_{t \to +\infty} \|z_i(t) - z_j(t)\| \neq 0 (i, j = 1, 2, \ldots, M)$ for $\phi_i \neq \phi_j$, then cluster synchronization manifold $S_U = \{[z_1^T(t), z_2^T(t), \ldots, z_m^T(t)]^T \mid z_i(t) \in \mathbb{R}^m, z_i(t) = z_j(t), \forall i, j \in U_w, i, j = 1, 2, \ldots, m, w = 1, 2, \ldots, r\}$ is globally stable. Namely, the cluster synchronization among the derivative CNNs (3.1) and the leader NNs (3.2) is achieved with q clusters.

Definition 3.3 (Average Impulsive Interval) ([36]) Let $T_\xi(s, t)$ be the impulsive times in the impulsive sequence $\xi = \{t_1, t_2, \ldots\}$, which fall in the time interval (t, s). If there exist two positive parameters T_0 and T_i satisfying

$$\frac{s-t}{T_i} - T_0 \le T_\xi(s,t) \le \frac{s-t}{T_i} + T_0, \quad \forall s \ge t \ge 0,$$

then for the impulsive sequence ξ, we say that the average impulsive interval is no larger than T_i.

Lemma 3.1 ([37]) *Suppose that there exists a positive function $V(t)$ satisfies the following impulsive differential inequalities*

$$\begin{cases} D^+V(t) \le -\rho V(t) + \zeta \sup_{\varrho \in [t-\tau, t]}\{V(\varrho)\}, \\ V(t) = \psi(t), \qquad t \in [t_0 - \tau, t_0] \end{cases}$$

for positive parameters ρ, ζ, then one may derive that $V(t) \le \bar{V}(t_0)e^{-\lambda(t-t_0)}$, $t \ge t_0$, where $\psi(t)$ is a piecewise continuous function and $\bar{V}(t_0) = \sup_{\varrho \in [t_0-\tau, t_0]}\{V(\varrho)\}$. In addition, $\lambda > 0$ is called the convergence velocity, which is the unique feasible solution to the equation $\lambda - \rho + \zeta e^{\lambda\tau} = 0$.

Lemma 3.2 ([38]) *If there exist a positive function $V(t, x)$ and two positive constants satisfying $\rho > \zeta \ge 0$ and conditions in Lemma 3.1, then we have*

$$\mathbb{E}[V(t, x(t))] \le \mathbb{E}[\bar{V}(t_0)]e^{-\lambda(t-t_0)} \tag{3.6}$$

for all $t > t_0 \ge 0$, where λ is defined in Lemma 3.1 and $\bar{V}(t_0) = \sup_{t \in [t_0-\tau, t_0]}$ $\{V(s, x(s))\}$.

Lemma 3.3 ([39]) *Consider a nonlinear impulsive system with stochastic disturbance*

$$\begin{cases} dY(t) = X(t, C(t))dt + \theta(t, Y(t))d\tilde{\omega}(t), \ t \ne t_k, \\ Y(t_k^+) - Y(t_k^-) = u_k(Y(t_k^-)), \ k \in \mathfrak{N}^+, \end{cases} \tag{3.7}$$

where $Y(t)$ is the state vector, $X(\cdot,\cdot), \theta(\cdot,\cdot), C(\cdot), u_k(\cdot)$ are constant functions. For the positive function $V(t, Y)$, if there exist two different nonlinear functions Λ, $\tilde{\psi}_k$ with $\Lambda(t, 0) = \tilde{\psi}_k(0) = 0$ satisfying that

(i) For any $t \ge t_0$, there exist two positive parameters b_1, b_2 that satisfy

$$b_1 \cdot \|Y(t)\| \le V(t, Y) \le b_2 \cdot \| Y(t) \|;$$

(ii) Define a continuous concave function as $\Lambda : \mathbb{R}^+ \times \mathbb{R}^+ \to \mathbb{R}$. Denote $V_Y(t, Y) = [\frac{\partial V(t,Y)}{\partial Y_1}, \frac{\partial V(t,Y)}{\partial Y_2}, \ldots, \frac{\partial V(t,Y)}{\partial Y_m}]$ and $V_{YY}(t, Y) = (\frac{\partial^2 V(t,Y)}{\partial Y_i \partial Y_j})_{m\times m}$. For the above conditions, there satisfies

$$\mathcal{L}V(t, Y) = V_t(t, Y) + V_Y(t, Y)\Lambda(t, Y) + \frac{1}{2}trace[\theta^T(t, Y)V_{YY}\theta(t, Y)];$$

(iii) *For the continuous concave functions* $\tilde{\psi}_k : \mathbb{R}^+ \to \mathbb{R}^+$, *if there holds the inequality*

$$V(t_k^+, Y(t_k^+)) \leq \tilde{\psi}_k(V(t_k^-, Y(t_k^-))).$$

Consequently, the solution of the following comparison system

$$\begin{cases} \dot{a}(t) = \phi(t, a(t)), \ t \neq t_k, \\ a(t_k^+) = \tilde{\psi}_k(a(t_k^-)), \ k \in \mathfrak{N}^+, \\ a(t_0) = \mathbb{E}[V(t_0, x_0)] \end{cases} \tag{3.8}$$

is exponentially stable, which is obvious that the exponential stability of the trivial solution of the nonlinear impulsive system with stochastic disturbance (3.7) is obtained, where $\mathbb{E}(\cdot)$ *means the mathematical expectation.*

Assumption 3.1 The two nonlinear system functions $f^1_{\phi_i}(\cdot)$, $f^2_{\phi_i}(\cdot)$ are supposed to satisfy the Lipschitz condition. Namely, there exist two related positive scalars $q^1_{\phi_i} > 0$ and $q^2_{\phi_i} > 0$, for any $a, b \in \mathbb{R}^m$, there satisfies

$$\|f^1_{\phi_i}(a) - f^1_{\phi_i}(b)\| \leq q^1_{\phi_i}\|a - b\|,$$
$$\|f^2_{\phi_i}(a) - f^2_{\phi_i}(b)\| \leq q^2_{\phi_i}\|a - b\|.$$

Assumption 3.2 Suppose that the matrix-valued function $h(\cdot, \cdot, \cdot, \cdot)$ is a uniformly Lipschitz function with the norm induced by the trace inner product on the matrices, that is

$$Trace[h^T(t, b, b_d, b_\tau)h(t, b, b_d, b_\tau)] \leq \|Wb\|^2 + \|W_d b_d\|^2 + \|W_\tau b_\tau\|^2$$

for any vectors $b(t), b_d(t), b_\tau(t) \in \mathbb{R}^m$, where W, W_d and W_τ are three given constant matrices.

3.3 Main Results

In this section, we will study the exponential cluster synchronization problem for the time-varying delay CNNs (3.4) with derivative coupling and stochastic disturbances. According to the theorems on the concept of average impulsive interval and the mathematical induction, sufficient conditions for cluster synchronization among the derivative CNNs (3.1) and the target NNs (3.2) will be acquired under impulsive pinning controller (3.3).

Theorem 3.1 *Suppose that Assumptions 3.1 and 3.2 hold and the average impulsive interval is no larger than* T_i *for the impulse time sequence* $\xi = \{t_1, t_2, \ldots\}$. *For the impulsive effects* ϕ *and* μ *in controller (3.3), if there are a matrix* $P =$

$diag\{p_1, p_2, \ldots, p_M\} > 0$ and positive parameters n, d, $q^1_{\phi_i}$, $q^2_{\phi_i}$, such that
***(i)** Matrix inequality*

$$\chi = \begin{pmatrix} c_1 L^s + nI_M - D & \frac{1}{2}c_2 Q \\ \frac{1}{2}c_2 Q^T & -dI_M \end{pmatrix} \leq 0; \tag{3.9}$$

***(ii)** For the given positive constant s, there holds*

$$\max\{b + ce^{\lambda\tau}, e^{\lambda\tau}\} \leq s, \tag{3.10}$$

where $\tau = \max\{\tau(t), \bar{\tau}(t), \check{\tau}(t)\}$, $a = \lambda_{\max}(I_M - c_3 G)(1+\phi)(1+\phi+\mu)$, $b = \lambda_{\max}(I_M - c_3 G)(1+\phi+\mu)\mu$, and λ is the unique feasible solution to the equation

$$\lambda - \rho + \sigma e^{\lambda\tau} = 0$$

with the conditions $\rho > \sigma \geq 0$, $\sigma = \zeta + \gamma$,

$$\zeta = \max_{1 \leq i \leq M} \{\lambda_{\max}(W_\tau^T W_\tau + q^2_{\phi_i} I_m)\},$$

$$\gamma = \max_{1 \leq i \leq M} \{\lambda_{\max}(W_{\bar{\tau}}^T W_{\bar{\tau}} + 2dI_m)\}$$

$$\rho = -\frac{\min_{1 \leq i \leq M}\{\lambda_{\min}(2A_{\phi_i} + B_{\phi_i} B_{\phi_i}^T + C_{\phi_i} C_{\phi_i}^T + q^1_{\phi_i} I_m + W^T W - 2nI_m)\}}{\lambda_{\max}(I_M - c_3 G)};$$

***(iii)** The exponential convergence velocity satisfies*

$$\frac{\ln s}{T_i} - \lambda < 0, \tag{3.11}$$

then the solutions of the impulsively controlled error NNs (3.4) is exponentially stable. That is, the exponential cluster synchronization among the derivative CNNs (3.1) and the target NNs (3.2) is eventually achieved by mean of the impulsive pinning control strategy (3.3).

Proof Construct the following Lyapunov function with considering the derivative coupling matrix G

$$V(t) = \frac{1}{2}\delta z(t)^T ((I_M - c_3 G) \otimes I_m)\delta z(t), \tag{3.12}$$

where $\delta y(t) = [\delta y_1^T(t), \delta z_2^T(t), \ldots, \delta z_M^T(t)]^T \in \mathbb{R}^{mM}$.

Firstly, By utilizing the Itô differential equations proposed in Lemma 3.3, for $t \in [t_{k-1}, t_k)$ with $k \in \mathfrak{N}^+$, calculating derivative $\mathcal{L}V(t)$ along the controlled error CNNs (3.4)

$$
\begin{aligned}
\mathcal{L}V(t) &= \delta z(t)^T((I_M - c_3 Q)\otimes I_m)\dot{\delta} z(t) + \frac{1}{2}\sum_{i=1}^{M} Trace(h(t, \delta z_i(t), \delta z_i(t-\tau(t)),\\
&\quad \delta z_i(t-\bar{\tau}(t))^T h(t, \delta z_i(t), \delta z_i(t-\tau(t)), \delta z_i(t-\bar{\tau}(t)))\\
&\le \sum_{i=1}^{M} \delta z_i(t)^T (A_{\phi_1}\delta z_i(t) + B_{\phi_i}\Phi_{\phi_i}(\delta z_i(t)) + c_1 \sum_{j=1}^{M} l_{ij}\delta z_j(t)\\
&\quad + C_{\phi_i}\Psi_{\phi_i}(\delta z_i(t-\tau(t))) + c_2 \sum_{j=1}^{M} q_{ij}\delta z_j(t-\bar{\tau}(t)) + c_3\sum_{j=1}^{M} g_{ij}\dot{\delta} z_j(t)\\
&\quad - p_i \delta z_i(t)) - c_3 \delta z(t)^T (Q\otimes I_m)\dot{\delta} z(t) + \frac{1}{2}\sum_{i=1}^{M}\delta z_i^T(t) W^T W \delta z_i(t)\\
&\quad + \frac{1}{2}\sum_{i=1}^{M}\delta z_i^T(t-\tau(t)) W_\tau^T W_\tau \delta z_i(t-\tau(t))\\
&\quad + \frac{1}{2}\sum_{i=1}^{M}\delta z_i^T(t-\bar{\tau}(t)) W_{\bar{\tau}}^T W_{\bar{\tau}} \delta z_i(t-\bar{\tau}(t))\\
&\le \min_{1\le i\le N}\{\lambda_{\min}(2A_{\phi_i} + B_{\phi_i}B_{\phi_i}^T + C_{\phi_i}C_{\phi_i}^T + q_{\phi_i}^1 I_m + W^T W - 2nI_m)\}\\
&\quad \times \frac{1}{2}\sum_{i=1}^{M}\delta z_i(t)^T\delta z_i(t) + \max_{1\le i\le M}\{\lambda_{\max}(q_{\phi_i}^2 I_m + W_\tau^T W_\tau)\}\\
&\quad \times \frac{1}{2}\delta z(t-\tau(t))^T((I_M - c_3 G)\otimes I_m)\delta z(t-\tau(t))\\
&\quad + \max_{1\le i\le M}\{\lambda_{\max}(2dI_m + W_{\bar{\tau}}^T W_{\bar{\tau}})\}\frac{1}{2}\delta z(t-\bar{\tau}(t))^T\\
&\quad \times ((I_M - c_3 G)\otimes I_m)\delta z(t-\bar{\tau}(t)) + \sum_{k=1}^{m}\Pi^T(t)\chi\Pi(t)\\
&\le -\rho V(t) + \zeta V(t-\tau(t)) + \gamma V(t-\bar{\tau}(t))\\
&\le -\rho V(t) + \sigma \sup_{\varrho\in[t-\bar{\tau},t]} V(\varrho)
\end{aligned}
\tag{3.13}
$$

where $\Pi(t) = (\tilde{\delta} z_k(t)^T, \tilde{\delta} z_k(t-\bar{\tau}(t))^T)^T$, $\tilde{\delta} z_k(t) = (\delta z_1^k(t), \delta z_2^k(t), \ldots, \delta z_M^k(t))^T$, $\bar{\tau} = \max\{\tau(t), \bar{\tau}(t)\}$, $\rho = -\frac{\eta}{\lambda_{\max}(I_M - c_3 Q)}$, $\eta = \min_{1\le i\le M}\{\lambda_{\min}(2A_{\phi_i}^s + B_{\phi_i}B_{\phi_i}^T + C_{\phi_i}C_{\phi_i}^T + q_{\phi_i}^1 I_m + W^T W - 2mI_m)\}$ $\zeta = \max_{1\le i\le M}\{\lambda_{\max}(W_\tau^T W_\tau + q_{\phi_i}^2 I_m)\}$, and $\gamma = \{\max_{1\le i\le M}\lambda_{\max}(W_{\bar{\tau}}^T W_{\bar{\tau}} + 2dI_m)\}$.

Secondly, for the impulsive time instant $t = t_k$, $k \in \mathfrak{N}^+$. In view of the impulsive equation in controlled error CNNs (3.4), we get

$$
\delta z_i(t_k^+) = (1+\phi)\delta z_i(t_k^-) + \mu\delta z_i(t_k^- - \check{\tau}(t_k^-)).
$$

By the fact that $\lambda_{\max}(I_M - c_3 G) \geq \lambda_{\min}(I_M - c_3 G) = 1$ in Remark 3.1, we have the following results

$$
\begin{aligned}
V(t_k^+) &= \frac{1}{2}\delta z^T(t_k^+)((I_M - c_3 G)\otimes I_m)\delta z(t_k^+) \\
&\leq \frac{1}{2}\lambda_{\max}(I_M - c_3 G)\delta z^T(t_k^+)\delta z(t_k^+) \\
&= \frac{1}{2}\lambda_{\max}(I_M - c_3 G)((1+\phi)\delta z(t_k^-) + \mu\delta z^T(t_k^- - \check{\tau}(t_k^-))) \\
&\quad \times ((1+\phi)\delta z(t_k^-) + \mu\delta z(t_k^- - \check{\tau}(t_k^-))) \\
&= \frac{1}{2}\lambda_{\max}(I_M - c_3 G)((1+\phi)^2\delta z^T(t_k^-)\delta z(t_k^-) \\
&\quad + 2(1+\phi)\mu\delta z^T(t_k^-)\delta z(t_k^- - \check{\tau}(t_k^-)) \\
&\quad + \mu^2\delta z^T(t_k^- - \check{\tau}(t_k^-))\delta z(t_k^- - \check{\tau}(t_k^-))) \\
&\leq \frac{1}{2}\lambda_{\max}(I_M - c_3 G)((1+\phi)^2\delta z^T(t_k^-)\delta z(t_k^-) \\
&\quad + (1+\phi)\mu\delta z^T(t_k^-)\delta z(t_k^-) \\
&\quad + (1+\phi)\mu\delta z^T(t_k^- - \check{\tau}(t_k^-))\delta z(t_k^- - \check{\tau}(t_k^-)) \\
&\quad + \mu^2\delta z^T(t_k^- - \check{\tau}(t_k^-))\delta z(t_k^- - \check{\tau}(t_k^-))) \\
&\leq \frac{1}{2}\frac{\lambda_{\max}(I_M - c_3 G)}{\lambda_{\min}(I_M - c_3 G)}((1+\phi)(1+\phi+\mu)\delta z^T(t_k^-) \\
&\quad \times ((I_M - c_3 G)\otimes I_m)\delta z(t_k^-) + (1+\phi+\mu)\mu\cdot \\
&\quad \delta z^T(t_k^- - \check{\tau}(t_k^-))((I_M - c_3 G)\otimes I_m)\delta z(t_k^- - \check{\tau}(t_k^-))) \\
&= \lambda_{\max}(I_M - c_3 G)((1+\phi)(1+\phi+\mu)V(t_k^-) \\
&\quad + (1+\phi+\mu)\mu V(t_k^- - \check{\tau}(t_k^-))) \\
&= bV(t_k^-) + cV(t_k^- - \check{\tau}(t_k^-)),
\end{aligned}
\tag{3.14}
$$

where $b = \lambda_{\max}(I_M - c_3 G)(1+\phi)(1+\phi+\mu)$, $c = \lambda_{\max}(I_M - c_3 G)\mu(1+\phi+\mu)$. For $k \in \mathfrak{N}^+$. Taking the mathematical expectation on both sides of above inequality gives

$$
\mathbb{E}[V(t_k^+)] \leq b\mathbb{E}[V(t_k^-)] + c\mathbb{E}[V(t_k^- - \check{\tau}(t_k^-))]. \tag{3.15}
$$

Therefore, based on Lemmas 3.1 and 3.2, for $\bar{V}(t_{k-1}) = \sup_{\varrho \in [t_{k-1}-\tau, t_{k-1}]} V(\varrho)$, considering the time interval $t \in [t_{k-1}, t_k)$ $k \in \mathfrak{N}^+$ obtains

$$
\mathbb{E}[V(t)] \leq \mathbb{E}[\bar{V}(t_{k-1})]e^{-\lambda(t-t_{k-1})}. \tag{3.16}
$$

Then, for the given parameter $s > 0$ and $t > t_0 > 0$, we shall derive

$$\mathbb{E}[V(t)] \leq s^{k-1}\mathbb{E}[\bar{V}(t_0)]e^{-\lambda(t-t_0)} \tag{3.17}$$

by applying the mathematical induction method.

Firstly, for the case that $t \in [t_0, t_1)$, If there exists a positive parameter s, then we have

$$\begin{aligned}\mathbb{E}[V(t)] &\leq \mathbb{E}[\bar{V}(t_0)]e^{-\lambda(t-t_0)} = s^0\mathbb{E}[\bar{V}(t_0)]e^{-\lambda(t-t_0)}\\ &= s^{k-1}\mathbb{E}[\bar{V}(t_0)]e^{-\lambda(t-t_0)}.\end{aligned}$$

Then, assume that the inequality (3.17) is true, when $k = \varrho$, i.e., $t \in [t_{\varrho-1}, t_\varrho)$, it follows form (3.10) in condition (ii) of Theorem 3.1 that

$$\begin{aligned}\mathbb{E}[V(t_\varrho)] &\leq b\mathbb{E}[V(t_\varrho^-)] + c\mathbb{E}[V(t_\varrho^- - \check{\tau}(t-\varrho))]\\ &\leq b \cdot s^{\varrho-1}\mathbb{E}[\bar{V}(t_0)] \cdot e^{-\lambda(t_\varrho - t_0)} + c \cdot s^{\varrho-1}\mathbb{E}[\bar{V}(t_0)]e^{-\lambda(t_\varrho - \check{\tau}(t-\varrho)-t_0)}\\ &\leq (b + ce^{\lambda\tau})s^{\varrho-1}\mathbb{E}[\bar{V}(t_0)]e^{-\lambda(t_\varrho - t_0)}\\ &\leq s^{\varrho}\mathbb{E}[\bar{V}(t_0)]e^{-\lambda(t_\varrho-t_0)}.\end{aligned}$$

Finally, we will discuss the situation that $\varrho + 1$. Namely, considering the time interval $t \in [t_\varrho, t_{\varrho+1})$ acquires

$$\begin{aligned}\mathbb{E}[V(t)] &\leq \mathbb{E}[\bar{V}(t_\varrho)]e^{-\lambda(t-t_\varrho)}\\ &= \sup_{\tilde{\eta}\in[t_\varrho-\tau, t_\varrho]} \mathbb{E}[V(\tilde{\eta})]e^{-\lambda(t-t_\varrho)}\\ &= \max\{\sup_{\tilde{\eta}\in[t_\varrho-\tau, t_\varrho)} \mathbb{E}[V(t)], \mathbb{E}[V(t_\varrho)]\}e^{-\lambda(t-t_\varrho)}\\ &\leq \max\{s^{\varrho-1}\mathbb{E}[\bar{V}(t_0)]e^{-\lambda(t_\varrho-\tau-t_0)}, s^{\varrho}\mathbb{E}[\bar{V}(t_0)]e^{-\lambda(t_\varrho-t_0)}\}e^{-\lambda(t-t_\varrho)}\\ &= \max\{e^{\lambda\tau}, s\}s^{\varrho-1}\mathbb{E}(\bar{V}(t_0))e^{-\lambda(t-t_0)}\\ &\leq s^{\varrho}\mathbb{E}[\bar{V}(t_0)]e^{-\lambda(t-t_0)}.\end{aligned}$$

Until now, it has been verified that for all time satisfy $t \in [t_{k-1}, t_k)$ $k \in \mathfrak{N}^+$, there is

$$\mathbb{E}[V(t)] \leq s^{k-1}\mathbb{E}[\bar{V}(t_0)]e^{-\lambda(t-t_0)}.$$

In Eq. (3.17), s is just a positive constant. Additionally, by using the concept of average impulsive interval, we would like to discuss the exponential cluster synchronization of the derivative CNNs according to the selection of the parameter $s > 0$.

(Case 1). If the positive parameter $s \in (0, 1)$ with $t \in [t_{k-1}, t_k)$ $k \in \mathfrak{N}^+$, we get

$$\begin{aligned}\mathbb{E}[V(t)] &\leq s^{k-1}\mathbb{E}[\bar{V}(t_0)]e^{-\lambda(t-t_0)}\\ &\leq s^{\frac{t-t_0}{T_i}-T_0}\mathbb{E}[\bar{V}(t_0)]e^{-\lambda(t-t_0)}\\ &= s^{-T_0}\mathbb{E}[\bar{V}(t_0)]s^{\frac{t-t_0}{T_i}}e^{-\lambda(t-t_0)}\\ &= s^{-T_0}\mathbb{E}[\bar{V}(t_0)]e^{(\frac{\ln s}{T_i}-\lambda)(t-t_0)}.\end{aligned} \tag{3.18}$$

(Case 2). If $s \in (1, +\infty)$, for $t \in [t_{k-1}, t_k)$ $k \in \mathfrak{N}^+$, it implies that

$$\begin{aligned}\mathbb{E}[V(t)] &\leq s^{k-1}\mathbb{E}[\bar{V}(t_0)]e^{-\lambda(t-t_0)}\\ &\leq s^{\frac{t-t_0}{T_i}+T_0}\mathbb{E}[\bar{V}(t_0)]e^{-\lambda(t-t_0)}\\ &= s^{T_0}\mathbb{E}[\bar{V}(t_0)]s^{\frac{t-t_0}{T_i}}e^{-\lambda(t-t_0)}\\ &= s^{T_0}\mathbb{E}[\bar{V}(t_0)]e^{(\frac{\ln s}{T_i}-\lambda)(t-t_0)}.\end{aligned} \tag{3.19}$$

(Case 3). Specially, when $s = 1$, for $t \in [t_{k-1}, t_k)$ $k \in \mathfrak{N}^+$, we have

$$\begin{aligned}\mathbb{E}[V(t)] &\leq s^{k-1}\mathbb{E}[\bar{V}(t_0)]e^{-\lambda(t-t_0)}\\ &= \mathbb{E}[\bar{V}(t_0)]e^{(\frac{\ln s}{T_i}-\lambda)(t-t_0)}.\end{aligned} \tag{3.20}$$

Through the analysis of the above three cases about the positive parameter s, three results (3.18), (3.19) and (3.20) are obtained, which related to the ultimate synchronization state. Therefore, for two positive parameters λ and M_0, the inequality (3.5) hold in Definition 3.1. Moreover, considering the specially constructed Lyapunov function (3.12), it gives

$$\|\delta z_i(t)\|^2 \leq \frac{2V(t)}{\lambda_{\min}(I_M - c_3 G)} = 2V(t). \tag{3.21}$$

Conducting mathematical expectations on above inequality obtains

$$\mathbb{E}[\|\delta z_i(t)\|^2] \leq M_0\mathbb{E}[\bar{V}(t_0)]e^{(\frac{\ln s}{T_i}-\lambda)(t-t_0)}. \tag{3.22}$$

Namely, the impulsively controlled error NNs (3.4) is exponentially stable with the convergence rate $\frac{1}{2}(\lambda - \frac{\ln s}{T_i})$. Additionally, the exponential cluster synchronization between the derivative CNNs (3.1) and the leader NN (3.2) are achieved in mean square in view of the impulsive pinning controller (3.3). Until now, we have finished the proof. □

Remark 3.2 In generally, for previous scholars who mainly focused on synchronization of complex networks investigated two main types of coupling methods, that is, general state coupling and delayed state coupling. However, considering network model with coupling only at time or/and time-delayed stated cannot entirely reflect the real situation of the information transmissions among the systems in the coupled networks. In some special engineering applications, based on couplings at time/time-delayed states, the derivative states should also be considered simultaneously when simulate the information transmission among different systems in coupled networks [32–34]. Additionally, the CNNs model (3.1) with three different types of coupling methods: general state coupling $c_1 \sum_{i=1}^{M} l_{ij}\Gamma z_j(t)$, delayed state coupling $c_2 \sum_{i=1}^{M} q_{ij}\Gamma y_j(t-\bar{\tau}(t))$ and derivative state coupling $c_3 \sum_{i=1}^{M} g_{ij}\Gamma \dot{z}_j(t)$, is considered in this chapter. Further, the general state $z_i(t)$ and delayed state $z_i(t-\bar{\tau}(t))$ are utilized to show the spatial position of the i-th NN, while the derivative state $\dot{z}_i(t)$ is applied to describes the velocity of information transmission among different NNs. Consequently, it can be concluded that the dynamical changing rate of the i-th NN is jointly determined by the dynamics of the itself i-th NN and the changing rate of its neighbor NNs.

Remark 3.3 It is obvious from $\max\{b + ce^{\lambda\tau}, e^{\lambda\tau}\} \leq s$ in condition (ii) of Theorem 3.1 that the positive parameter s is not only directly related to the impulsive effects ϕ and μ but also to the upper bounds τ of time-delayed. It should be noted that the smaller impulsive effects ϕ and μ will result in the smaller s. Based on the techniques of the average impulsive interval, Case 1, Case 2 and Case 3 are conducted before, respectively, to prove the definition of exponential synchronization. In some previous works like [18, 24, 27], when designing the impulsive controllers, the positive and negative impulses for network synchronizing have been studied, respectively. Therefore, the synchronization of complex networks needs to satisfy the different conditions, which will lead to higher requirements on control performance and system parameters. However, in this chapter, no matter the impulsive effects ϕ and μ are positive or negative to the final synchronization of the complex networks, the uniform synchronization conditions are obtained because of the effectiveness of the classification discussion methods on the positive constant s in three situations and the introduction of the average impulsive interval. In other word, as long as all conditions in Theorem 3.1 are satisfied, the cluster synchronization among the derivative CNNs (3.1) and the target NNs (3.2) could be acquired under impulsive pinning controller (3.3).

Remark 3.4 It should be noted that the impulsive controller consists of two parts: the impulsive control part $u_{0,i}(t)$ and the pinning negative feedback part $u_{1,i}(t)$. Correspondingly, two different parts play different roles in network synchronization. As we all know, the impulsive control is a kind of superior discontinuous control strategy which only actively works at each impulsive instant in a very sparse sequence of time, thus compared with the continuous control methods, the control cost is greatly reduced when the same control effect is achieved. It is obvious that in this chapter, the impulsive effects ϕ and μ are closely related to the final results of the networks [17–20, 22, 25, 26]. On the hand, if two impulsive effects ϕ and μ work enough to synchronize the networks, then the feedback control strength p_i in controller $u_{1,i}$ could be some small constants or even take zero. On the other hand, when the impulsive effects ϕ or μ not enough to synchronize, or even hamper the final synchronization, the negative feedback control term $-p_i\delta z_i(t)$ would work efficiently to counteract the adverse effects resulted from the insufficient impulses. In addition, from the point of negative feedback inputs in $u_{1,i}$ of controller $u_i(t)$, the term $-p_i\delta z_i(t)$ is used to synchronize all NNs in the ϕ_i-th cluster, while the remaining items are used to weaken the mutual influence among clusters at the intersection NNs.

Remark 3.5 In fact, by adjusting two positive parameters T_I and T_0, the impulsive times of the impulsive sequence ξ with impulsive interval (t, s) could be estimated because the concept of the average impulsive interval was introduced in Definition 3.3 in this chapter. In the most previous works focused on impulsive control problems, like [17, 19, 21], the impulsive interval T_i was briefly defined as $T_i = \min_k\{t_k - t_{k-1}\}$ or $T_i = \max_k\{t_k - t_{k-1}\}$, which finally leads to higher requirements on controller parameters and causes the waste of resources. To discuses the impulsive control issues, the concept of average impulsive interval has been introduced [24, 33, 36, 38]. For the impulsive interval related constant T_i, we have $T_i \in [\min_k\{t_k - t_{k-1}\}, \max_k\{t_k - t_{k-1}\}]$. Different from previous works [17, 19, 21], in this chapter, according to the concept of average impulsive intervals, the evaluation on T_i would efficiently reduce the conservatism in conditions and requirements on system parameters and control strengths.

Remark 3.6 In this chapter, we only discuss the time invariant impulsive effects ϕ and μ in order to investigate simplicity. However, for some special engineering applications, the impulsive effects may be changed at different impulsive instants, due to the difference the control subjects and the synchronization targets. In other word, impulsive effects ϕ and μ designed in the impulsive pinning controller could be replaced by ϕ_k and μ_k, and thus, condition (ii) in Theorem 3.1 should be replaced by the following condition:
(ii') For the upper bounds of the three time-varying delays τ, there holds

$$\max\{a_k + b_k e^{\lambda\tau}, e^{\lambda\tau}\} \le s, \tag{3.23}$$

where $a_k = \lambda_{\max}(I_M - c_3 G)(1 + \phi_k)(1 + \phi_k + \mu_k)$, $b_k = \lambda_{\max}(I_M - c_3 G)(1 + \phi_k + \mu_k)\mu_k$, $k \in \mathfrak{N}^+$, and the parameter $\lambda > 0$ is the only solution to the following equation

$$\lambda - \rho + \sigma e^{\lambda\tau} = 0$$

with $\rho > \sigma \ge 0$, $\sigma = \zeta + \gamma$,

$$\zeta = \max_{1\le i\le M}\{\lambda_{\max}(W_\tau^T W_\tau + q_{\phi_i}^2 I_m)\},$$

$$\gamma = \max_{1\le i\le M}\{\lambda_{\max}(W_{\bar{\tau}}^T W_{\bar{\tau}} + 2d I_m)\},$$

$$\rho = -\frac{\min_{1\le i\le M}\{\lambda_{\min}(2A_{\phi_i}^s + B_{\phi_i}B_{\phi_i}^T + C_{\phi_i}C_{\phi_i}^T + q_{\phi_i}^1 I_m + W^T W - 2n I_m)\}}{\lambda_{\max}(I_M - c_3 G)}.$$

3.4 Numerical Simulation

In this simulation, we consider the NN with stochastic disturbance [40]

$$\begin{aligned} dz(t) = [&A_q z(t) + B_q f_q^1(z(t)) + C_q f_q^2(z(t-\tau(t))) \\ &+ I(t)]dt + \tilde{h}(t, z(t), z(t-\tau(t)), z(t-\bar{\tau}(t)))d\tilde{\omega}(t), \end{aligned} \tag{3.24}$$

where

$$A_q = \begin{bmatrix} -1 & 0 \\ 0 & -1 \end{bmatrix}, \quad B_q = \begin{bmatrix} 2.0 & -0.1 \\ -5.0 & 4.5 \end{bmatrix},$$

$$C_q = \begin{bmatrix} -1.5 & -0.1 \\ -0.2 & -4 \end{bmatrix}, \quad I(t) = \begin{bmatrix} 0 \\ 0 \end{bmatrix},$$

$z(t) = [z(t)^1, z(t)^2]^T$ nonlinear functions $f_1^1(z) = f_1^2(z) = f_2^1(z) = f_2^2(z) = \tanh(z)$, time-varying delay $\tau(t) = 1 + 0.1\sin(0.1t)$, q stands for the number of clusters, the stochastic matrix

$$\tilde{h}(z(t), z(t-\tau(t)), z(t-\bar{\tau}(t))) = 0.5 \cdot \|z(t)\| I_2,$$

$\tilde{\omega}(t)$ is an 2-dimensional Brownian motion. Select the initial values as $z^1(t) = 0.2$, $z^2(t) = 0.3$, for $t \in [-0.2, 0]$, we present the phase graph of the described NN (3.24) in the first figure of Fig. 3.1.

In this numerical example, consider multiple CNNs with six NNs, which are separated into two different clusters $\Lambda_1 = \{1, 2, 3\}$ and $\Lambda_2 = \{4, 5, 6\}$ (see the second picture of Fig. 3.1) It is obvious from the cluster-tree structure of the CNNs, the third NN in the first cluster is directly connected with the forth NN in the second cluster. Therefore, the impulsive pinning controller (3.3) will be imposed on the third and the forth NNs, then we have corresponding negative feedback control inputs $u_{1,3}(t)$ and $u_{1,4}(t)$. Let the coupling matrices be

$$L = \begin{pmatrix} -1 & 1 & 0 & 0 & 0 & 0 \\ 0 & -2 & 2 & 0 & 0 & 0 \\ 3 & 0 & -3 & 0 & 0 & 0 \\ 0 & 0 & 4 & -5 & 1 & 0 \\ 0 & 0 & 0 & 0 & -5 & 5 \\ 0 & 0 & 0 & 6 & 0 & -6 \end{pmatrix},$$

$$Q = \begin{pmatrix} -5 & 1 & 0 & 1 & 1 & 2 \\ 1 & -4 & 2 & 0 & 1 & 0 \\ 0 & 3 & -6 & 1 & 0 & 2 \\ 2 & 0 & 0 & -4 & 1 & 1 \\ 1 & 1 & 2 & 0 & -6 & 2 \\ 0 & 2 & 1 & 0 & 1 & -4 \end{pmatrix},$$

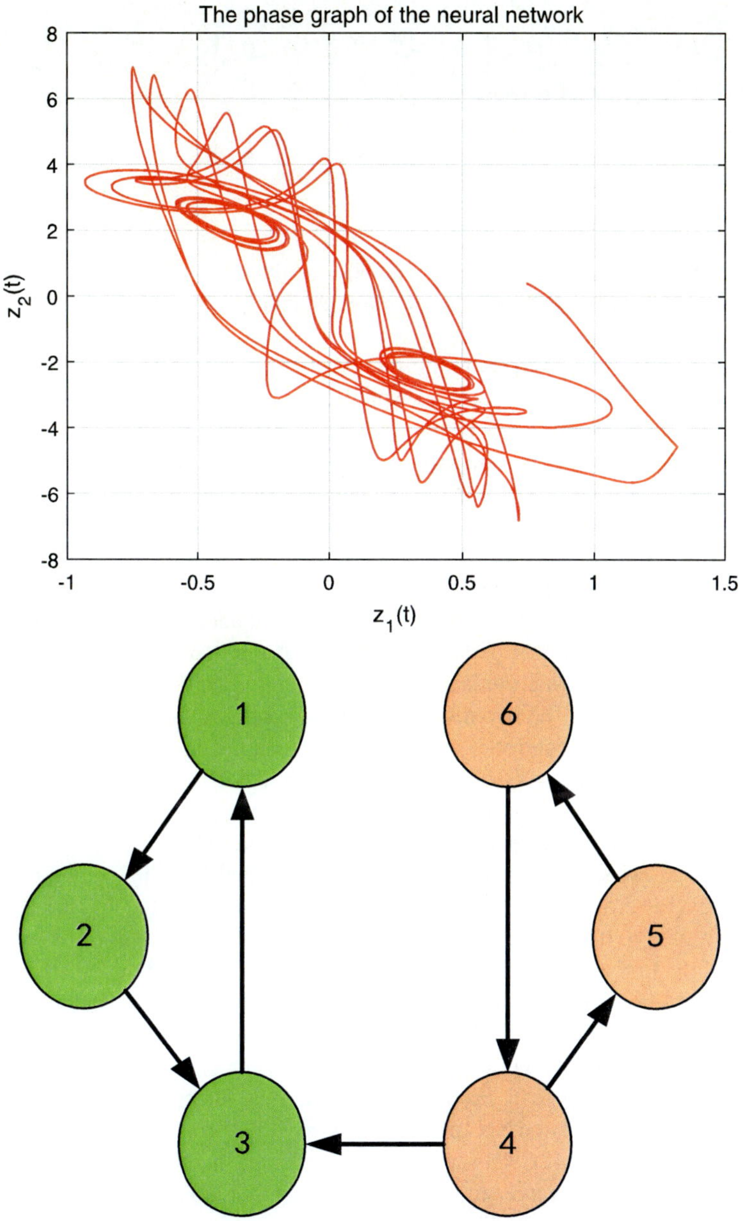

Fig. 3.1 The phase graph of the NNs (27) and the topology structure of the derivative CNNs

$$G = \begin{pmatrix} -3 & 1 & 0 & 2 & 0 & 0 \\ 1 & -2 & 1 & 0 & 0 & 0 \\ 0 & 2 & -3 & 1 & 0 & 0 \\ 1 & 1 & 0 & -5 & 2 & 1 \\ 1 & 2 & 2 & 0 & -6 & 1 \\ 0 & 0 & 2 & 1 & 1 & -4 \end{pmatrix},$$

the feedback control strength matrix $P = \mathrm{diag}\{0, 0, 0.5, 0.5, 0, 0\}$, coupling strengths $c_1 = 0.5, c_2 = 0.2, c_3 = 0.1$, and coupling time-varying delay $\bar{\tau}(t) = 0.2\sin(0.2t)$. Then by some simple calculations, it could be obtained that $\lambda_{\max}(I_6 - c_3 G) = 1.5212$.

For designing the impulsive pinning controller $u_{0i}(t)$ in (3.3), we consider the following nonuniform distribution of impulses, which comes form [36]. Let $\xi = \{x, 2x, \ldots, (T_0 - 1)x, T_0 T_i, T_0 T_i + x, T_0 T_i + 2x, \ldots, T_0 T_i + (T_0 - 1)x, 2T_0 T_i, \ldots\}$. That means $t_k - t_{k-1} = x$ if $\mathrm{mod}(k, T_0) \neq 0$ and $T_0(T_i - x) + x$, if $\mathrm{mod}(k, T_0) = 0$, where $0 \leq x \leq T_i$. To illustrate the effectiveness of impulsive control, we take the average impulsive interval is no more than $T_i = 0.02$, and the free adjust parameters $T_0 = 2, x = 0.01$.

For the given system parameters, we get $\rho = 58.7001, \zeta = 1, \gamma = 3.5800$, and it satisfies $\rho > \zeta + \gamma = 4.5800 > 0$. Let $\check{\tau}(t) = 0.2$. In addition, one could calculate the solution to the nonlinear parameter equation $\lambda - \rho + \sigma e^{\lambda\tau} = 0$ is $\lambda = 2.2828$. To realize the cluster synchronization between the derivative CNNs (3.1) and the leader NN (3.2), select the two different impulsive effects as $\phi = 0.5$, $\mu = 0.1$ for the impulsive pinning controller (3.3). Then, we acquire $b = 3.6509$, $c = 0.2434$. Moreover, one obtains $\max\{b + ce^{\lambda\tau}, e^{\lambda\tau}\} = \max\{6.6492, 12.3182\} = 12.3182 \leq s = 12.32$ from condition (ii) in Theorem 3.1. Additionally, from the inequality (14) in condition (iii) of Theorem 3.1, we can get $\frac{\ln s}{T_i} - \lambda = -1.0272 < 0$.

In Fig. 3.2a, we show the state evolution curves of the CNNs in the first cluster. It obvious that the error states $e_i^j(t)(i = 1, 2, 3, j = 1, 2)$ among the first three NNs in the first cluster $z_i(t)(i = 1, 2, 3)$ and the synchronization objective $s_{\phi_i}(t)(\phi_i = 1, i = 1, 2, 3)$ of the first cluster approach to zero within 20 s. In addition, the states of the three NNs in the second cluster are plotted in Fig. 3.3a. The error states $e_i^j(t)(i = 4, 5, 6, j = 1, 2)$ among the NNs $z_i(t)(i = 4, 5, 6)$ in the second cluster and the synchronization objective $s_{\phi_i}(\phi_i = 2, i = 4, 5, 6)$ of the second cluster approach to zero within 16 s in Fig. 3.3b. From above the figures, the error states between the CNNs and the target isolated NN approach to zero within a certain period of time. That is, the synchronization of CNNs are achieved within each cluster. Further, we could get the exponential convergence rate of the synchronization is $|\frac{\ln s}{T_i} - \lambda| = 3.8740$ under the definition in Eq. (3.11).

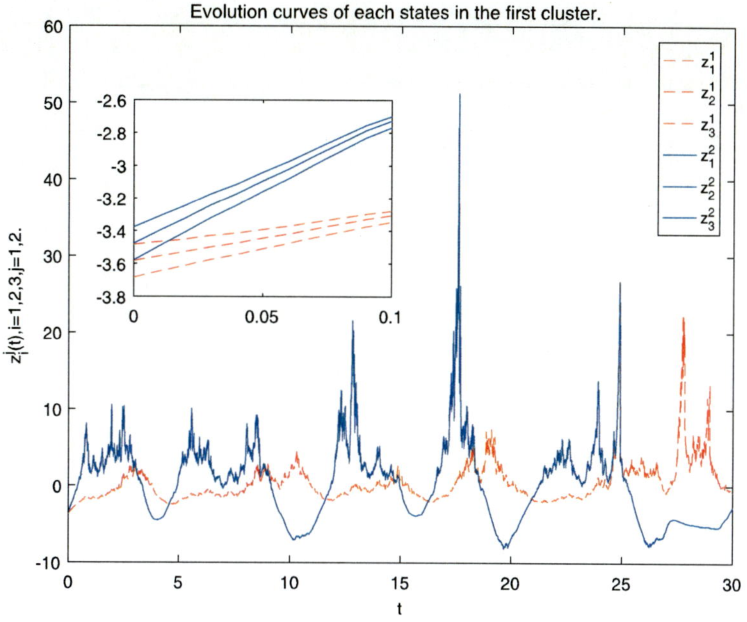

(a)

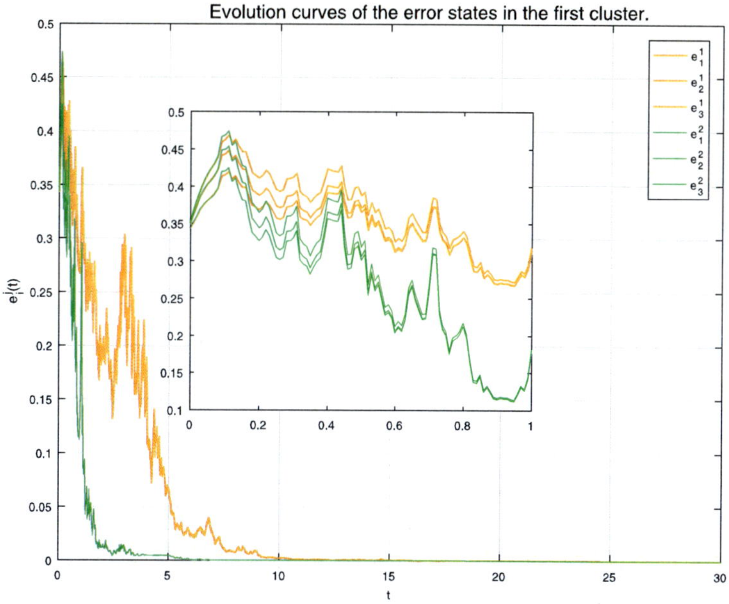

(b)

Fig. 3.2 **a** Two states evolution curves of the NNs in the first cluster. **b** The evolution curves of the error states in the first cluster

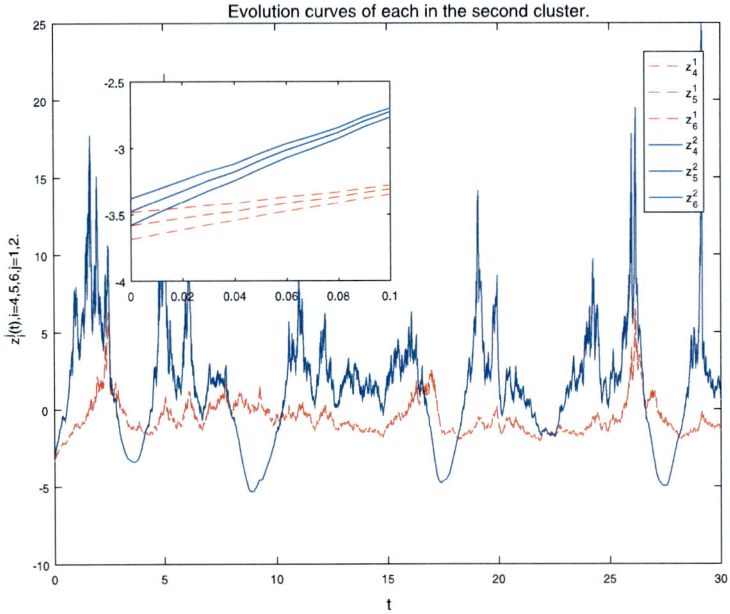

(a)

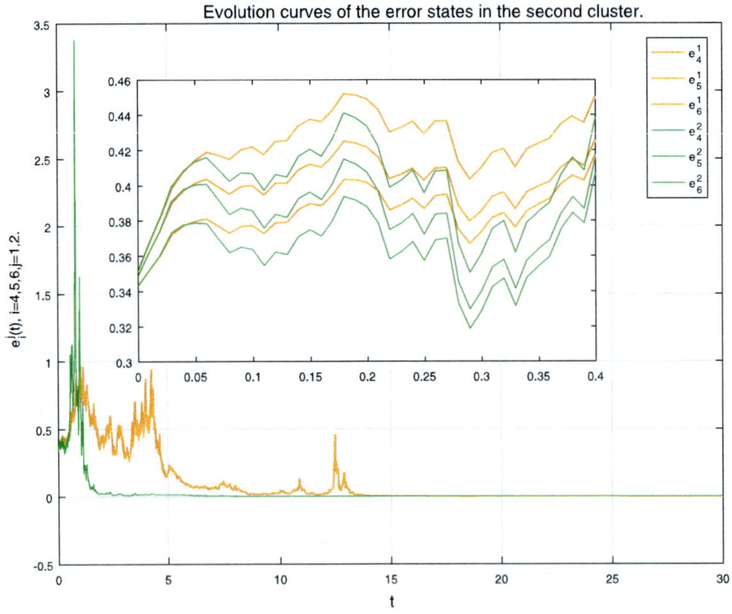

(b)

Fig. 3.3 **a** Two states evolution curves of the NNs in the second cluster. **b** The evolution curves of the error states in the second cluster

3.5 Conclusion

In this chapter, the global and exponential cluster synchronization of CNNs with multiple time-varying delays and stochastic disturbance has been investigated. For the network modeling, three different types of coupling mechanisms including general state coupling, delayed state coupling and state derivative coupling have been considered. A novel impulsive pinning feedback control protocol has been proposed with fully considering the cluster-tree topology structures of the networks and efficiently control the NNs directly connected in different clusters. Based on the concept of average impulsive interval, the mathematical induction method and the Lyapunov stability theorem, sufficient conditions have been derived that ensure the realization of the global and exponential cluster synchronization on the derivative CNNs. Simultaneously, the exponential convergence rate of the cluster synchronization has been precisely evaluated. Finally, a numerical simulation has been implemented to prove the correction of the theoretical analysis and control schemes.

References

1. Strogatz, S. H.: Exploring complex networks. Nature. **410**, 268–276 (2001)
2. Huang, C. X., Cao, J. D.: Stochastic dynamics of non-autonomous Cohen-Grossberg neural networks. Abstract and Applied Analysis. **2011**, 1–17 (2011)
3. Liu, H., Wan, H. B., Tse, C. K., Lv, J. H.: An encryption scheme based on synchronization of two-layered complex dynamical networks. IEEE Transactions on Circuits and Systems I: Regular Papers. **63**, 2010–2021 (2016)
4. Huang, C. X., Cao, J., Wang, P.: Attractor and boundedness of switched stochastic Cohen-Grossberg neural networks. Discrete Dynamics in Nature and Society. **2016**, 1–19 (2016)
5. Benson, A. R., Gieich, D. F., Leskovec, J.: Higher-order organization of complex networks. Science. **353**, 163–166 (2016)
6. Li, C. G., Chen, G. R.: Synchronization in general complex dynamical networks with coupling delays. Physica A: Statistical Mechanics and its Applications. **343**, 263–278 (2004)
7. Huang, C. X., Yang, X. S., He, Y. G., Huang, L. H.: Stability of stochastic Reaction-Diffusion recurrent neural networks with unbounded distributed delays. Discrete Dynamics in Nature and Society. **2011**, 1–16 (2011)
8. Song, X. M., Yan,, X. H.: Linear quadratic Gaussian control for linear time-delay systems. IET Control Theory and Applications. **8**, 375–383 (2014)
9. Yan, X. H., Song, X. M., Wang, X.: Global output-feedback stabilization for nonlinear time-delay systems with unknown control coefficients. International Journal of Control, Automation. **16**, 1550–1557 (2018)
10. Huang, C. X., Zhang, H., Huang, L. H.: Almost periodicity analysis for a delayed Nicholson's blowflies model with nonlinear density-dependent mortality term. Communications on Pure and Applied Analysis. **18**, 3337–3349 (2019)
11. Wu, Z. G., Shi, P., Su, H. Y., Chu, J.: Stochastic synchronization of Markovian jump neural networks with time-varying delay using sampled-data. IEEE Transactions on Cybernetics. **43**, 1796–1806 (2013)
12. Cai, C. X., Wang, Z. D., Xu, J., Liu, X. H., Alsaadi, F. E.: An integrated approach to global synchronization and state estimation for nonlinear singularly perturbed complex networks. IEEE Transactions on Cybernetics. **45**, 1597–1609 (2015)

13. Huang, C. X., Zhang, H., Cao, J. D., Hu, H. J.: Stability and Hopf bifurcation of a delayed prey predator model with disease in the predator. International Journal of Bifurcation and Chaos. **29**, 1–23 (2019)
14. Wang, J. L., Wu, H. N., Huang, T. W., Ren, S. Y., Wu, J. G.: Pinning control for synchronization of coupled reaction-diffusion neural networks with directed topologies. IEEE Transactions on Cybernetics. **46**, 1109–1120 (2016)
15. Liu, X. Y., Ho, Daniel W. C., Xie, C. L.: Prespecified-time cluster synchronization of complex networks via a smooth control approach. IEEE Transactions on Cybernetics. **50**, 1771–1775 (2018)
16. Ma, Q. C., Qin, J. H., Zheng, W. X., Kang, Y.: Output group synchronization for networks of heterogeneous linear systems under internal model principle. IEEE Transactions on Circuits and Systems I: Regular Papers. **65**, 1684–1695 (2018)
17. Lu, J. Q., Wang, Z. D., Cao, J. D., Ho, D. W. C., Kurths, J.: Pinning impulsive stabilization of nonlinear dynamical networks with time-varying delay. International Journal of Bifurcation and Chaos. **22**, 1–12 (2012)
18. Chen, Y., Yu, W. W., Li, F. F., Feng, S. S.: Synchronization of complex networks with impulsive control and disconnected topology. IEEE Transactions on Circuits and Systems II: Express Briefs. **60**, 292–296 (2013)
19. Zhang, W. B., Tang, Y., Miao, Q. Y., Fang, J. A.: Synchronization of stochastic dynamical networks under impulsive control with time delays. IEEE Transactions on Neural Networks and Learning Systems. **25**, 1758–1768 (2014)
20. Yang, Z. C., Xu, D. Y.: Stability analysis and design of impulsive control systems with time delay. IEEE Transactions on Automatic Control. **52**, 1448–1454 (2007)
21. Hu, J. Q., Liang, J. L., Cao, J. D.: Synchronization of hybrid-coupled heterogeneous networks: pinning control and impulsive control schemes. Journal of the Franklin Institute. **351**, 2600–2622 (2014)
22. He, W. L., Chen, G. R., Han, Q. L., Qian, F.: Network-based leader-following consensus of nonlinear multi-agent systems via distributed impulsive control. Information Science. **380**, 145–158 (2017)
23. Guan, Z. H., Liu, Z. W., Feng, G., Wang, Y. W.: Synchronization of complex dynamical networks with time-varying delays via impulsive distributed control. IEEE Transactions on Circuits and Systems I: Regular Papers. **57**, 2182–2195 (2010)
24. Tang, Z., Park, J. H., Feng, J. W.: Impulsive effects on quasi-synchronization of neural networks with parameter mismatches and time-varying delay. IEEE Transactions on Neural Networks and Learning Systems. **29**, 908–919 (2018)
25. He, W. L., Qian, F., Han, Q. L., Cao, J. D.: Synchronization error estimation and controller design for delayed Lur'e systems with parameter mismatches. IEEE Transactions on Neural Networks and Learning Systems. **23**, 1551–1563 (2012)
26. He, W. L., Qian, F., Lam, J., Chen, G. R., Han, Q. L., Kurths, J.: Quasi-synchronization of heterogeneous dynamic networks via distributed impulsive control: Error estimation, optimization and design. Automatica. **62**, 249–262 (2015)
27. Xu, W. Y., Ho, D. W. C.: Clustered event-triggered consensus analysis: An impulsive framework. IEEE Transactions on Industrial Electronics. **63**, 7133–7143, (2016)
28. Yang, X. S., Ho, D. W. C., Lu, J. Q., Song, Q.: Finite-time cluster synchronization of T-S fuzzy complex networks with discontinuous subsystems and random coupling delays. IEEE Transactions on Fuzzy Systems. **23**, 2302–2316 (2012)
29. Tang, Z., Park, J. H., Lee, T. H., Feng, J. W.: Random adaptive control for cluster synchronization of complex networks with distinct communities. International Journal of Adaptive Control and Signal Processing. **30**, 534–549 (2016)
30. Tang, Z., Park, J. H., Shen,, H.: Finite-time cluster synchronization of Lur'e networks: a nonsmooth approach. IEEE Transactions on Systems, Man, and Cybernetics: Systems. **48**, 1213–1224 (2018)
31. Zhou, L. L., Wang, C. H., Du, S. C., Zhou, L.: Cluster synchronization on multiple nonlinearly coupled dynamical subnetworks of complex networks with nonidentical nodes. IEEE Transactions on Neural Networks and Learning Systems. **28**, 570–583 (2017)

32. Xu, Y. H., Zhou, W. N., Fang, J. A., Xie, C. R., Tong, D. B.: Finite-time synchronization of the complex dynamical network with non-derivative and derivative coupling. Neurocomputing. **173**, 1356–1361 (2016)
33. Zheng, S.: Pinning and impulsive synchronization control of complex dynamical networks with non-derivative and derivative coupling. Journal of the Franklin Institute. **354**, 6341–6363 (2017)
34. Deng, L. P., Wu, Z. Y., Wu, Q. C.: Pinning synchronization of complex network with non-derivative and derivative coupling. Nonlinear Dynamics. **73**, 775–782 (2013)
35. Wu, W., Zhou, W. J., Chen, T. P.: Cluster synchronization of linearly coupled complex networks under pinning control. IEEE Transactions on Circuits and Systems I: Regular Papers. **56**, 829–839 (2009)
36. Lu, J. Q., Ho, D. W. C., Cao, J. D.: A unified synchronization criterion for impulsive dynamical networks. Automatica. **46**, 1215–1221 (2010)
37. Halanay, A.: Differential equations: stability, oscillations, time lags. Academic Press. New York, USA (1996)
38. Tang, Z., Park, J. H., Lee, T. H., Feng, J. W.: Mean square exponential synchronization for impulsive coupled neural networks with time-varying delays and stochastic disturbances. Complexity. **21**, 190–202 (2016)
39. Liu, B.: Stability of solutions for stochastic impulsive systems via comparison approach. IEEE Transactions on Automatic Control. **53**, 2128–2133 (2008)
40. Yang, X. S., Cao, J. D.: Stochastic synchronization of coupled neural networks with intermittent control. Physical Letters A. **373**, 3259–3272 (2009)

Chapter 4
Adaptively Synchronize the Derivative Coupled CDNs with Proportional Delay

4.1 Introduction

A complex dynamic network (CDN) is generally considered to be the basic structure of a complex system, and any complex system can be thought of as a network of interactions between units or individuals. The importance of a complex network in describing its complexity is determined by its structure. The study of complex networks is a method of studying complex systems, paying great attention to the dynamics of each system and the topology structure of the interacting system, i.e. the communication between different systems in the network [1]. In consequence, the analysis and discussion on complex networks becomes the basis for understanding the characteristics and functions of complex systems.

The relationship between complex network topology and synchronization has become one of the most pressing topics in researching the complex networks in recent years. The aim of synchronization is to force all or part of the systems in complex networks to some desired objectives by regulating the system parameters, the coupling strengths or imposing some control inputs [2]. Furthermore, only a few dynamical networks could achieve synchronization via adjusting system parameters and coupling strengths on their own [2] not only in practical networks but also in artificial networks. However, for realizing the synchronization of the majority of the complex networks, it is highly necessary to input the external forces [3–9]. In recent years, based on the results of numerous researches, the problem of strengthening the capability in realizing the synchronization for complex networks has drawn the attention from many scholars. Hence, for the purpose of dealing with various synchronization problems, like reducing the communication congestion and the spread of disease, diverse control strategies have been proposed, for instance, intermittent control, adaptive, impulsive control and slide mode control [10–15]. Besides, it is universally recognized that, as a special discontinuous control strategy, impulsive control could effectively reduce the control costs in contrast to the general

Z. Tang et al., *Impulsive Synchronization of Complex Dynamical Networks*,
https://doi.org/10.1007/978-981-16-5383-4_4

continuous control protocols, for the control action only exerted on the system in the discrete pulse instant [16, 17].

In fact, due to certain reasons, such as the unexpected altering of the running surroundings and the jam of information communication channels [18–21], the complex dynamical networks are subject to high frequency of uncertainties. As one of the most common uncertainties that is hard to avoid, time delay, which would seriously affects the performance of the networks, and in serious cases, even cause the each dynamical system to be unstable. Obviously, in recently, the time delay problems have been fully discussed with generally considering two classical types delays, time invariant delay τ and time-varying delay $\tau(t)$ [22, 23]. For example, the corresponding synchronization criteria for time-delay complex networks have been derived in references [24–28]. Nevertheless, the simple forms of time-delay like τ or $\tau(t)$ are not enough in modeling some complicated situations in practical applications. Therefore, the complex networks with proportional delay always thought as the quality of service models because of some special characteristics, such as controllable and predictable. Consequently, in order to simulate the data exchange among the systems in complex dynamical networks, as one of unbounded delays, proportional delay should be taken into consideration [29–34]. Furthermore, due to the numerous parallel communication channels with various axon sizes and lengths, the modelling of complex network is frequently expected to consider some proportional delays while in a certain time interval. Specifically, $x(pt)$ is always used to indicate the proportional time delay, where $p \in (0, 1)$ represents the proportional delay rate. Moreover, $x(pt) = x(t - \tau(t))$ could be derived from the definition of $\tau(t) = t - pt$, where $\tau(t) \to +\infty$ when $t \to +\infty$. In other words, the unboundary and monotonically increasing are two main differences between the proportional time delay and general time delay. It is obviously that, the analysis difficulty is largely increased for the synchronization on complex dynamical networks comparing with the case where the bounded delay issues. These reasons could directly lead to the improperly applying the traditional methods to solve the issues, such as generalized Lyapunov functions and some mathematical inequalities. For example, the finite-time stability problem for the fuzzy cellular neural networks with proportional time-varying delays was discussed in [32], where the finite-time stability theory and some inequality techniques were introduced. Whereafter, based on the matrix theory and Lyapunov functional method, the asymptotic stability of equilibrium point of a kind of cellular neural networks with multiple proportional time-varying delays was investigated in [34]. It should be noted that the state $x(qt)$ with proportional delay in a dynamical system is time proportionally shifted in the step of qt with the proportional delay ratio q instead of linearly transformed by normal time delay $t - \tau(t)$ in above mentioned works studying proportional delays. Through constructing an variable transformation $x_i(t) = y_i(e^t)$, this problem could be settled, while the substantial increase would occur in complexity and difficulty within calculations. In this sense, the problem of synchronization on complex networks becomes challengeable while with proportional delay.

On the other side, it is known to us that a complex network is normally constituted by a number of dynamical systems, therefore, there exist information interactions among different systems, which means there is a strong coupling among each other. It is certainly that some general coupling patterns, such as general state coupling, delayed state coupling and derivative coupling, have been considered until now in describing the information exchange in complex networks. In particular, general state coupling and delay state coupling represent the interactions of location information between each system, while for the derivative coupling, it means the velocity information of the state transmission of each system is interactive [35–40]. Furthermore, we can get that both of the state information of the current system and its adjacent systems jointly determine the state transition speed of the complex network. For instance, in [35], the issue of synchronization on the complex dynamical networks with derivative and non-derivative coupling was fully studied, as well as the effective finite-time synchronization criteria were derived. In addition, researchers fully investigated the synchronization problems of bipartite dynamical networks through utilizing Lyapunov stability theory and adaptive strategy in [39], where the distributed delays and the nonlinear derivative coupling were considered simultaneously.

To the best of authors' knowledge, until now, the adaptive synchronization problem of derivative coupled complex dynamical networks with proportional delay, has been rarely discussed. Due to its valuableness and potential in both theory and practical applications, it deserves to be studied. In this paper, the global synchronization of such complex dynamical network with derivative couplings is discussed by introducing an adaptive impulsive pinning protocol. The distinctive characteristics of this paper could be summarized as follows: (i) The exponential synchronization of a class of complex networks with derivative coupling is studied, meanwhile, designing an extended Lyapunov function corresponding to derivative coupling; (ii) The proportional delay is introduced while considering more complicated situations in practical applications. Meanwhile, the impulsive comparison principle is taken as an analytical method for dealing with proportional delay instead of introducing some variable transformations like $x_i(t) = y_i(e^t)$ in [29–34]; (iii) Sufficient exponential synchronization criteria of the complex dynamical network with derivative coupling under pinning control and impulsive control strategies are obtained by jointly applying the extended parameters variation formula and the definition of average impulsive interval; (iv) Suitable control gains of the pinning term for exponentially synchronize the complex network are derived by applying the designed impulsive control protocol attached valid adaptive updating laws. Furthermore, the convergence rate to achieve exponential synchronization is accurately estimated. Besides, the effectiveness of the designed control schemes and the main results are verified through a numerical simulation. Last but not least, the concept of impulsive distance is first introduced, then the dynamic balance between the impulsive effects and the feedback control gains is further explained by these examples, which provides a method for designing control protocol.

The organization of the rest paper is as follows. The derivative coupled complex network model with proportional delay and some preliminariesis are given in

Sect. 4.2. In Sect. 4.3, the synchronization criteria and exponential convergence rate are obtained of this kind of derivative coupled complex networks. Furthermore, the correctness of the theoretical results is verified in Sect. 4.4. And we make the conclusion of this paper in Sect. 4.5.

4.2 Problem Formulation

4.2.1 Model Description

Consider the following derivative coupled complex dynamical networks with proportional delay

$$\dot{x}_i(t) = f(x_i(t)) + c_1\sum_{j=1}^{N} h_{ij}\Gamma\dot{x}_j(t) + c_2\sum_{j=1}^{N} b_{ij}\Gamma x_j(pt), \tag{4.1}$$

where the n-dimensional state vector of the i-th system is denoted by $x_i(t) = [x_i^1(t), x_i^2(t), \ldots, x_i^n(t)]^T \in \mathbb{R}^n$; Dynamic function $f : \mathbb{R}^n \to \mathbb{R}^n$ is the memoryless nonlinear vector-valued function which is continuous differentiable on $\mathbb{R}$; The coupling strengths c_1, c_2 are all positive constants and the inner connection matrix is $\Gamma = \text{diag}\{r_1, r_2, \ldots, r_n\} \in \mathbb{R}^{n\times n}$, $\Gamma > 0$. Specially, it is assumed that $\Gamma = I_n$ in this paper; $H = [h_{ij}]_{N\times N}$ represents the derivative coupled matrix with the derivative term $\dot{x}_j(t)$, and the proportional delay coupled matrix with the proportional delay term $x_j(pt)$ is denoted by $b = [b_{ij}]_{N\times N}$. Two matrices are all decided by the topology structures of the network, and the two coupling matrices are supposed to satisfy diffusive condition, that is, $h_{ii} = -\sum_{j\neq ij=1}^{N} h_{ij} = -\sum_{j\neq ij=1}^{N} h_{ji}$ and $b_{ii} = -\sum_{j\neq ij=1}^{N} b_{ij} = -\sum_{j\neq ij=1}^{N} b_{ji}$. In addition, $h_{ji} = h_{ij} > 0\,(b_{ji} = b_{ij} > 0)$ if there is a connection between the i-th system and the j-th system for $i \neq j$, and $h_{ij} = 0\ (b_{ij} = 0)$ otherwise. The factor $p \in (0, 1)$ is a constant which involves the history time. In particular, the dynamics of the i-th system at time t in the complex network model (4.1) is determined by the states $x_j(t)$ for $j = 1, 2, \ldots, N$, and the states $x_j(pt)$ at history time pt which is proportional to current time instant t with a constant rate p. Therefore, the constant p is regarded as proportional delay. And thus, we could get that $pt = t - \tau(t)$, where $\tau(t) = (1 - p)t \geq 0$ but $\tau(t) \to \infty$ when $t \to \infty$. From this point, proportional delay could be regarded as a class of unbounded time-varying delay.

Remark 4.1 The eigenvalues of the matrix H can be arranged as $0 = \lambda_1 \geq \lambda_2 \geq \cdots \geq \lambda_N$, based on the above definition. Furthermore, we can get matrix $I_N - c_1 H$ is a positive definite matrix with the minimum eigenvalues $\lambda_{\min}(I_N - c_1 G) = 1$ and $\lambda_{\max}(I_N - c_1 H) \geq 1$. These derivations are prepared for the construction of the Lyapunov functional below.

Consider the following isolated solution of dynamic equation (4.1)

$$\dot{s}(t) = f(s(t)), \tag{4.2}$$

where $s(t) = [s^1(t), s^2(t), \ldots, s^n(t)]^T \in \mathbb{R}^n$. The error vector is defined with $e_i(t) = x_i(t) - s(t)$ ($e_i(t) = [e_i^1(t), e_i^2(t), \ldots, e_i^n(t)]^T \in \mathbb{R}^n$) for $i = 1, 2, \ldots, N$. Then the error network by imposing controller $u_i(t)$ on the complex network (4.1) can be obtained

$$\begin{aligned}\dot{e}_i(t) = {} & f(x_i(t)) - f(s(t)) + c_1 \sum_{j=1}^{N} h_{ij}\dot{e}_j(t) \\ & + c_2 \sum_{j=1}^{N} b_{ij} e_j(pt) + u_i(t), \ i = 1, 2, \ldots, N.\end{aligned} \tag{4.3}$$

The following adaptive impulsive control strategy is elaborately designed in order to achieve the adaptively exponential synchronization between the derivative coupled complex networks (4.1) and the dynamical system (4.2)

$$u_i(t) = -d_i(t)e_i(t) + \rho \sum_{k=1}^{+\infty} e_i(t)\delta(t - t_k), \tag{4.4}$$

where $d_i(t) \geq 0$ ($i = 1, 2, \ldots, N$) denotes time-varying adaptive feedback control gains that can be suitably selected by the complex network (4.1), and the impulsive effect is represented by ρ, nonlinear function $\delta(\cdot)$ is called the Dirac function. Then for the impulse sequence $\zeta = \{t_1, t_2, \ldots\}$, which satisfies $t_{k-1} < t_k$ and $\lim_{k\to+\infty} t_k = +\infty$ for $k = 1, 2, \ldots$.

For the purpose of obtaining the appropriate control strengths, the adaptive updating laws of $d_i(t)$ are designed as

$$\dot{d}_i(t) = \varepsilon_i e_i^T(t)e_i(t), \ i = 1, 2, \ldots, N, \tag{4.5}$$

where ε_i ($i = 1, 2, \ldots, N$) is a positive constant to ensure negative feedback.

Therefore, the following controlled error complex dynamical network with proportional delay is obtained

$$\begin{cases} \dot{e}_i(t) = f(x_i(t)) - f(s(t)) - d_i(t)e_i(t) \\ \qquad +c_1 \sum_{j=1}^{N} h_{ij}\dot{e}_j(t) + c_2 \sum_{j=1}^{N} b_{ij} e_j(qt), \ t \neq t_k, \\ \Delta e_i(t_k) = e_i(t_k^+) - e_i(t_k^-) = \rho e_i(t_k^-), \ t = t_k, \end{cases} \tag{4.6}$$

where $k = 1, 2, \ldots$ is positive integer. Hereafter, denote $k \in \mathcal{N} = 1, 2, \ldots$. Then $e_i(t)$ is assumed right-hand continuous at $t = t_k$, $k \in \mathcal{N}$ throughout this paper, which means $e_i(t_k) = e_i(t_k^+) = \lim_{\epsilon\to 0^+} e_i(t_k + \epsilon)$, $e_i(t_k^-) = \lim_{\epsilon\to 0^-} e_i(t_k + \epsilon)$. Thus, the solutions of (4.6) are piecewise right-hand continuous functions with discontinuities at $t = t_k$ for $k \in \mathcal{N}$.

Rewriting the above controlled error complex networks (4.6) into the following compact form by applying the Kronecker product gives

$$\begin{cases} \dot{e}(t) = F(X(t)) - F(S(t)) - (D(t) \otimes I_n)e(t) \\ \qquad +c_1(H \otimes I_n)\dot{e}(t) + c_2 B \otimes I_n)e(qt),\ t \neq t_k, \\ \Delta e(t_k) = e(t_k^+) - e(t_k^-) = \rho e(t_k^-),\ t = t_k, \end{cases} \tag{4.7}$$

where $e(t) = (e_1^T(t), e_2^T(t), \ldots, e_N^T(t))^T$, $F(X(t)) = (f(x_1(t))^T, f(x_2(t))^T, \ldots, f(x_N(t))^T)^T$, $F(s(t)) = (f(s(t))^T, f(s(t))^T, \ldots, f(s(t))^T)^T$. And $D(t) = \text{diag}\{d_1(t), d_2(t), \ldots, d_N(t)\}$.

4.2.2 Preliminaries

Some necessary definitions, lemmas and assumptions are given in advance for achieving the adaptively exponential synchronization of the derivative coupled complex networks with proportional delays.

Definition 4.1 For any initial values $\phi_i(t)$ and $t > \widetilde{T} > 0$, if there exist three non-negative constants $\widetilde{T}, \widetilde{\lambda}, \widetilde{\zeta}$ satisfy the following formula

$$\|e_i(t)\| = \|x_i(t) - s(t)\| \leq \widetilde{\zeta} f(\phi_i(t)) e^{-\widetilde{\lambda} t}, \quad i = 1, 2, \ldots, N,$$

then the exponential synchronization is successfully achieved, where function $f(\cdot) : \mathbb{R}^n \to \mathbb{R}$. Furthermore, $\widetilde{\lambda}$ could be regarded as the exponential convergence rate during the synchronization procedure.

Definition 4.2 ([13]) (*Average Impulsive Interval*) The average impulsive interval of the impulsive sequence $\zeta = \{t_1, t_2, \ldots\}$ is equal to N_a if there exist positive integer N_0 and positive number N_a, such that

$$\frac{T-t}{N_a} - N_0 \leq N_\zeta(T, t) \leq \frac{T-t}{N_a} + N_0, \quad \forall T \geq t \geq 0, \tag{4.8}$$

where $N_\zeta(T, t)$ denotes the number of impulsive times of the impulsive sequence ζ on the interval (t, T).

Lemma 4.1 ([14]) (Comparison Principle) *Denote a function class as $\mathcal{K}$ where function $\phi(\cdot) \in \mathcal{K}$ are piecewise continuous. Further, select two functions $a(t)$ and $b(t)$ from $PC(1)$, where $PC(1) = \{\phi | \phi : [-h, \infty) \to \mathbb{R}^l\}$ and $0 \leq h(t) \leq h\tau$. Suppose there are some constants p, q and ς satisfying the following conditions*

$$\begin{cases} D^+ a(t) \leq p a(t) + q a(t - h(t)),\ t \neq t_k, k \in \mathcal{N}, \\ a(t_k) \leq \varsigma a(t_k^-),\ k \in \mathcal{N}, \end{cases}$$

$$\begin{cases} D^+b(t) > pb(t) + qb(t-h(t)), \ t \neq t_k, k \in \mathcal{N}, \\ b(t_k) = \varsigma b(t_k^-), \ k \in \mathcal{N}, \end{cases}$$

then, $a(t) \leq b(t)$ *for* $t > 0$ *can be derived.*

Assumption 4.1 The nonlinear dynamical function $f(\cdot)$ is supposed to satisfy globally Lipschitz condition, that is, there exists a positive constant m satisfying the following inequality

$$\|f(u) - f(v)\| \leq m\|u - v\|, \quad u, v \in \mathbb{R}^n.$$

4.3 Main Results

In this section, sufficient conditions for the achievement of the adaptively exponential synchronization between the derivative coupled complex dynamical networks (4.1) and the dynamical system (4.2) are obtained according to impulsive control and adaptive pinning control protocols have been developed.

Theorem 4.1 *Suppose that the dynamical function* $f(\cdot)$ *of controlled error complex network (4.7) satisfies the Assumption 4.1, and the average impulsive interval is less than* N_a*. For the impulsive effect* ρ *in controller (4.4), define* $\delta = (1+\rho)^2\lambda_{\max}(I_N - c_1 H)$*, which is assumed* $\delta \geq 1$*, and if there exist a control gain matrix* $D = diag\{d_1, d_2, \ldots, d_N\} > 0$*, scalars* $\alpha > 0$*,* $\beta > 0$*,* $m > 0$*, such that*
(i) *The matrix inequality satisfies*

$$\frac{1}{2}c_2(B \otimes I_n)(B \otimes I_n)^T + \frac{1}{2}\alpha(I_N - c_1 H) \otimes I_n + hI_{Nn} - D \otimes I_n < 0; \qquad (4.9)$$

(ii) *The matrix inequality satisfies*

$$I_{Nn} - \beta(I_N - c_1 H) \otimes I_n < 0; \qquad (4.10)$$

(iii) *There exists a positive constant* $\lambda \in (0, -\upsilon)$ *such that*

$$\begin{cases} \upsilon + \lambda p + \beta\delta^{N_0} < 0, \\ \lambda(\upsilon + \lambda p + p\beta\delta^{N_0}) + \upsilon(\upsilon + \lambda p + \beta\delta^{N_0}) > 0, \\ \lambda^2(\upsilon + \lambda p + p\beta\delta^{N_0}) > \upsilon^2(\upsilon + \lambda p + \beta\delta^{N_0}), \end{cases} \qquad (4.11)$$

where $\upsilon = \frac{\ln\delta}{N_a} - \alpha < 0$*, then the exponential stability can be achieved of the controlled error complex networks (4.6) or (4.7) with the convergence velocity* $\frac{\lambda}{2}$*, that is, the adaptive exponential synchronization between the derivative coupled complex dynamical networks (4.1) and the leader system (4.2) is obtained under the impulsive adaptive pinning controller (4.4) jointly by the adaptive updating laws (4.5).*

Proof Choose the Lyapunov function as

$$V(t) = \frac{1}{2}e^T(t)((I_N - c_1 H) \otimes I_n)e(t) + \sum_{i=1}^{N} \frac{1}{2\varepsilon_i}(d_i(t) - d_i)^2, \tag{4.12}$$

where $d_i (i = 1, 2, \ldots, N)$ is nonnegative constant.

On one hand, for $t = t_k, k \in \mathcal{N}$, we have the following results based on the error derivative coupled complex dynamical network (4.7)

$$\begin{aligned}
V(t_k^+) &= \frac{1}{2}e(t_k^+)^T((I_N - c_1 H) \otimes I_n)e(t_k^+) + \sum_{i=1}^{N} \frac{1}{2\varepsilon_i}(d_i(t_k^+) - d_i)^2 \\
&\le \frac{1}{2}\lambda_{\max}(I_N - c_1 H)e(t_k^+)^T e(t_k^+) + \sum_{i=1}^{N} \frac{1}{2\eta_i}(d_i(t_k^+) - d_i)^2 \\
&= \frac{1}{2}\lambda_{\max}(I_N - c_1 H)(1+\rho)^2 e(t_k^-)^T e(t_k^-) + \sum_{i=1}^{N} \frac{1}{2\varepsilon_i}(d_i(t_k) - d_i)^2 \\
&\le \frac{\lambda_{\max}(I_N - c_1 H)}{2\lambda_{\min}(I_N - c_1 H)}(1+\rho)^2 e(t_k^-)^T((I_N - c_1 H) \otimes I_n)e(t_k^-) \\
&\quad + \sum_{i=1}^{N} \frac{1}{2\varepsilon_i}(d_i(t_k) - d_i)^2 \\
&= (1+\rho)^2 \lambda_{\max}(I_N - c_1 G)(\frac{1}{2}e(t_k^-)^T((I_N - c_1 H) \otimes I_n)e(t_k^-) \\
&\quad + \sum_{i=1}^{N} \frac{1}{2\varepsilon_i}(d_i(t_k) - d_i)^2) \\
&= (1+\rho)^2 \lambda_{\max}(I_N - c_1 H)V(t_k^-) \\
&= \delta V(t_k^-).
\end{aligned} \tag{4.13}$$

On the other hand, for $t \in [t_{k-1}, t_k)$ with $k \in \mathcal{N}$, computing the derivative of (4.12) along the controlled error complex network (4.7) obtains

$$\begin{aligned}
\dot{V}(t) &= e(t)^T((I_N - c_1 H)\otimes I_n)\dot{e}(t) + \sum_{i=1}^{N}\frac{1}{\varepsilon_i}(d_i(t) - d_i)\dot{d}_i(t) \\
&= e(t)^T\dot{e}(t) - e(t)^T(c_1 H\otimes I_n)\dot{e}(t) + \sum_{i=1}^{N}(d_i(t_k^+) - d_i)e_i(t)^T e_i(t) \\
&= e(t)^T(F(X(t)) - F(S(t)) + (c_1 H\otimes I_n)\dot{e}(t) + (c_2 B\otimes I_n)e(pt)) \\
&\quad - \sum_{i=1}^{N} d_i(t)e_i(t)^T e_i(t) - e(t)^T(c_1 H\otimes I_n)\dot{e}(t) \\
&\quad + \sum_{i=1}^{N}(d_i(t) - d_i)e_i(t)^T e_i(t) \\
&\le me(t)^T e(t) + e(t)^T(c_2 B\otimes I_n)e(pt)) - e(t)^T(D\otimes I_n)e(t) \\
&\le me(t)^T e(t) + \frac{1}{2}c_2 e(t)^T(B\otimes I_n)(B\otimes I_n)^T e(t)) \\
&\quad + \frac{1}{2}e(pt)^T e(pt) - e(t)^T(D\otimes I_n)e(t) \\
&= e(t)^T(mI_{Nn} + \frac{1}{2}c_2(B\otimes I_n)(B^T\otimes I_n)(D\otimes I_n) + \frac{1}{2}\alpha(I_N - c_1 H)\otimes I_n)e(t) \\
&\quad + \frac{1}{2}e(pt)^T(I_{Nn} - \beta(I_N - c_1 H)\otimes I_n))e(pt) \\
&\quad - \frac{1}{2}\alpha e(t)^T(I_N - c_1 H)\otimes I_n)e(t) + \frac{1}{2}\beta e(pt)^T((I_N - c_1 H)\otimes I_n)e(pt) \\
&\le -\frac{1}{2}\alpha e(t)^T(I_N - c_1 H)\otimes I_n)e(t) + \frac{1}{2}\beta e(pt)^T((I_N - c_1 H)\otimes I_n)e(pt) \\
&= -\alpha V(t) + \beta V(pt).
\end{aligned} \tag{4.14}$$

By considering the inequities (4.13) and (4.14), the following impulsive comparison system can be obtained for any $Q > 0$

$$\begin{cases}
\dot{y}(t) = -\alpha\cdot y(t) + \beta\cdot y(pt) + Q,\ t\neq t_k, k\in\mathcal{N}, \\
y(t_k^+) = \delta y(t_k^-),\ t = t_k, k\in\mathcal{N}, \\
y(0) = \sum_{i=1}^{N}\|\psi_i(0)\|^2 = \|\psi(0)\|.
\end{cases} \tag{4.15}$$

Obviously, it has a unique solution $y(t)$. According to the comparison principle in Lemma 4.1, it obtains that $V(t)\le y(t)$ for $\forall\, t > 0$. In view of the formula for parameters variation [41], the following equation with proportional delay term $y(pt)$ can be given

$$y(t) = H(t,0)y(0) + \int_0^t H(t,\theta)(\beta y(p\theta) + Q)d\theta, \tag{4.16}$$

where $H(t,\theta)$ is the Cauchy matrix of the following impulsive system

$$\begin{cases} \dot{y}(t) = -\alpha \cdot y(t), & t \neq t_k, k \in \mathcal{N}, \\ y(t_k^+) = \delta y(t_k^-), & t = t_k, k \in \mathcal{N}. \end{cases}$$

Since $\delta \geq 1$, one could calculate the Cauchy matrix $H(t, \theta)$ as follows according to the concept of the average impulsive interval

$$\begin{aligned} H(t,\theta) &= e^{-\alpha(t-\theta)} \prod_{\theta \leq t_k \leq t} \rho \\ &\leq e^{-\alpha(t-\theta)} \delta^{N_\varsigma(t,\theta)} \\ &\leq e^{-\alpha(t-\theta)} \delta^{\frac{t-\theta}{N_a}+N_0} \\ &\leq \delta^{N_0} e^{(\frac{\ln\delta}{N_a}-\alpha)(t-\theta)}. \end{aligned} \tag{4.17}$$

Furthermore, substituting (4.17) into the integral equation (4.16) gives that

$$\begin{aligned} y(t) &= H(t,0)y(0) + \int_0^t H(t,\theta)(\beta y(p\theta) + Q)d\theta \\ &\leq \delta^{N_0} \sum_{i=1}^{N} \|\psi_i(0)\|^2 e^{(\frac{\ln\delta}{N_a}-\alpha)t} + \int_0^t \delta^{N_0} e^{(\frac{\ln\delta}{N_a}-\alpha)(t-\theta)} (\beta y(p\theta) + Q)d\theta \\ &= \delta_0 e^{\upsilon t} + \int_0^t e^{\upsilon(t-\theta)} \delta^{N_0} (\beta y(p\theta) + Q)d\theta, \end{aligned} \tag{4.18}$$

where $\delta_0 = \delta^{N_0} \sum_{i=1}^{N} \|\psi_i(0)\|^2$ and $\upsilon = \frac{\ln\delta}{N_a} - \alpha$.

In the following, it is necessary that $\upsilon < 0$, for the purpose of deriving the exponential estimation of $y(t)$ by the mathematical contradiction method with constructing some inequalities based on analysis in inequality (4.18).

In view of the above discussions, we will prove that there holds the following results if there exists λ which satisfies (4.11)

$$y(t) < \delta_0 e^{-\lambda t} - \frac{\delta^{N_0} Q}{\rho + \beta\delta^{N_0}} \quad \forall t \geq 0. \tag{4.19}$$

Firstly, if $t = 0$ we have

$$y(0) < \delta_0 - \frac{\delta^{N_0} Q}{\upsilon + \beta\delta^{N_0}}.$$

Next, the validation of the conclusion (4.19) will be proved through the mathematical proof method: contradiction. If (4.19) is not true for all $t > 0$, that is, at least, there is a time instant $T > 0$ such that

$$y(T) \geq \delta_0 e^{-\lambda T} - \frac{\delta^{N_0} Q}{\upsilon + \beta\delta^{N_0}}, \quad T > t > 0, \tag{4.20}$$

but for $0 < t < T$, there still holds (4.19). Therefore, according to (4.17) and (4.18), (4.19), the following inequality can be obtained

$$
\begin{aligned}
y(T) &\leq \delta_0 e^{\upsilon T} + \int_0^T e^{\upsilon(T-\theta)} \delta^{N_0}(\beta y(p\theta) + Q) d\theta \\
&< \delta_0 e^{\upsilon T} + \int_0^T e^{\upsilon(T-\theta)} \beta \delta^{N_0}(\delta_0 e^{-\lambda p\theta} - \frac{\delta^{N_0} Q}{\upsilon + \beta\delta^{N_0}}) d\theta \\
&\quad + \int_0^T e^{\upsilon(T-\theta)} \delta^{N_0} Q d\theta \\
&= \delta_0 e^{\upsilon T} + \beta\delta^{N_0}\delta_0 e^{\upsilon T} \int_0^T e^{-(\upsilon+\lambda p)\theta} d\theta \\
&\quad + (\beta\delta^{N_0}\sigma + \delta^{N_0} Q) e^{\upsilon T} \int_0^T e^{\upsilon\theta} d\theta \\
&= \delta_0 e^{\upsilon T} - \frac{\beta\delta^{N_0}\delta_0}{\upsilon + \lambda p} e^{\upsilon T}(e^{-(\upsilon+\lambda p)T} - 1) - \frac{\delta^{N_0}(\beta\upsilon + Q)}{\upsilon} e^{\upsilon T}(e^{-\upsilon T} - 1) \\
&= \delta_0 e^{\upsilon T} - \frac{\beta\delta_0\delta^{N_0}}{\upsilon + \lambda p}(e^{-\lambda p T} - e^{\upsilon T}) - \frac{\delta^{N_0}(\beta\sigma + Q)}{\upsilon}(1 - e^{\upsilon T}), \qquad (4.21)
\end{aligned}
$$

where $\sigma = -\frac{\delta^{N_0} Q}{\upsilon + \beta\delta^{N_0}}$. Based on the definition of σ, it further gives

$$
y(t) \leq \delta_0 e^{\upsilon T} - \frac{\beta\delta_0\delta^{N_0}}{\upsilon + \lambda p}(e^{-\lambda p T} - e^{\upsilon T}) - \sigma(1 - e^{\upsilon T}). \qquad (4.22)
$$

In order to manifest the contradiction between the results (4.22) and (4.19), firstly, a parameter function is defined

$$
\Phi(t) = \upsilon(\upsilon + \lambda p + \beta\delta^{N_0}) e^{\upsilon t} + \lambda(\upsilon + \lambda p + p\beta\delta^{N_0}) e^{-\lambda t}, \qquad (4.23)
$$

then we take

$$
\dot{\Phi}(t) = \upsilon^2(\upsilon + \lambda p + \beta\delta^{N_0}) e^{\upsilon t} - \lambda^2(\upsilon + \lambda p + p\beta\delta^{N_0}) e^{-\lambda t}. \qquad (4.24)
$$

Particularly, $\dot{\Phi}(t) = 0$ if t satisfies

$$
t = T_\Phi = \frac{1}{\upsilon + \lambda} \ln \frac{\lambda^2((\upsilon + \lambda p) + p\beta\delta^{N_0})}{\rho^2((\rho + \lambda p) + \beta\delta^{N_0})},
$$

and from the condition (4.11), it gives

$$
0 < \frac{\lambda^2((\upsilon + \lambda p) + p\beta\delta^{N_0})}{\upsilon^2((\upsilon + \lambda p) + \beta\delta^{N_0})} < 1.
$$

As a consequence, it means $\Phi(t)$ is a increasing function when $0 < t < T_s$ for $0 < t < T_\Phi$, $\dot{\Phi}(t) > 0$. And $\dot{\Phi}(t) < 0$ if $t > T_s$, it denotes that $\Phi(t)$ is a decreasing function for $t > T_s$. In addition, by condition (4.11), there holds

$$\Phi(0) = \upsilon((\upsilon + \lambda p) + \beta\delta^{N_0}) + \lambda((\upsilon + \lambda p) + p\beta\delta^{N_0}) > 0.$$

Furthermore, $\lim_{t\to+\infty} \Phi(t) = 0$ for $\rho < 0$ can be gotten according to the definition of (4.23). As a consequence, it derives that $\Phi(t) \geq 0$ for $t \geq 0$.

Thinking that $\rho + \lambda p < 0$ and $\Phi(t) \geq 0$ gives

$$\upsilon(1 + \frac{\beta\delta^{N_0}}{\upsilon + \lambda p})e^{\upsilon t} + \lambda(1 + \frac{q\beta\delta^{N_0}}{\upsilon + \lambda p})e^{-\lambda t} \leq 0. \tag{4.25}$$

Considering the proportional delay factor $0 < q < 1$ and the constant $0 < \lambda < -\upsilon$, it is easily get $e^{-\lambda t} < e^{-\lambda pt}$. Then the following inequality holds

$$\upsilon(1 + \frac{\beta\delta^{N_0}}{\upsilon + \lambda p})e^{\upsilon t} + \lambda\frac{p\beta\delta^{N_0}}{\upsilon + \lambda p}e^{-\lambda pt} + \lambda e^{-\lambda t} \leq 0. \tag{4.26}$$

Define another parameter equation $\Xi(t)$ as

$$\Xi(t) = (1 + \frac{\beta\delta^{N_0}}{\upsilon + \lambda p})\, e^{\upsilon t} - \frac{\beta\delta^{N_0}}{\upsilon + \lambda p}\, e^{-\lambda pt} - e^{-\lambda t}, \tag{4.27}$$

where there hold that $\Xi(0) = 0$ and $\dot{\Xi}(t) \leq 0$ from (4.25). It further implies that $\Xi(t)$ is a monotone decreasing function with initial value $\Xi(0) = 0$. Consequently, $\Xi(t) \leq \Xi(0) = 0$ could be held, that is

$$\begin{aligned}
&(1 + \frac{\beta\delta^{N_0}}{\upsilon + \lambda p})\, e^{\upsilon t} - \frac{\beta\delta^{N_0}}{\upsilon + \lambda p}\, e^{-\lambda pt} \\
&= e^{\upsilon t} + \frac{\beta\delta^{N_0}}{\upsilon + \lambda p}\,(e^{\upsilon t} - e^{-\lambda pt}) \leq e^{-\lambda t}, t \geq 0.
\end{aligned} \tag{4.28}$$

Jointly taking the inequalities (4.22) and (4.28) gives

$$\begin{aligned}
y(T) &\leq \delta_0(1 + \frac{\beta\delta^{N_0}}{\upsilon + \lambda p})\, e^{\upsilon T} - \frac{\delta_0\beta\delta^{N_0}}{\upsilon + \lambda p}\, e^{-\lambda pT} + \sigma(1 - e^{\upsilon T}) \\
&\leq \delta_0\, e^{-\lambda T} + \sigma(1 - e^{\upsilon T}) \\
&< \delta_0\, e^{-\lambda T} + \sigma = \delta_0 e^{-\lambda T} - \frac{\delta^{N_0} Q}{\upsilon + \beta\delta^{N_0}},
\end{aligned} \tag{4.29}$$

it results in a contradiction to the assumption (4.20). Then the assumption (4.19) is true for $t \geq 0$. Furthermore, the following inequality could be obtain in view of the comparison principle

$$0 < V(t) \leq y(t) < \delta_0 e^{-\lambda t} - \frac{\delta^{N_0} Q}{\upsilon + \beta \delta^{N_0}}.$$

Letting $\varepsilon \to 0$ obtains

$$\|e(t)\| \leq \sqrt{\frac{2\delta^{N_0}}{\lambda_{\min}(I_N - c_1 H)}} e^{-\frac{\lambda}{2}t}.$$

Above all, it has been proved that the controlled error complex networks (4.6) is exponential stable with the convergence rate $\frac{\lambda}{2}$. That is, the adaptive exponential synchronization between the derivative coupled complex networks (4.1) and the dynamical system (4.2) is ultimately obtained under the adaptive impulsive pinning controller (4.4) and the adaptive updating laws (4.5). Until now, the proof of the theorem is totally completed. □

Remark 4.2 When modeling practical engineering based on complex network theory, coupling is generally considered for realizing the information exchange between various systems, and its forms include state coupling $\sum_{i=1}^{N} h_{ij}\Gamma x_j(t)$, time-delayed coupling $\sum_{i=1}^{N} h_{ij}\Gamma x_j(t - \tau(t))$ and derivative coupling $\sum_{i=1}^{N} h_{ij}\Gamma \dot{x}_j(t)$. In the existing researches on delay coupling, the two conditions like $0 \leq \tau(t) \leq \tau$ or $0 < \dot{\tau}(t) \leq \bar{\tau} < 1$ are always used to restrict time-delayed coupling to make it bounded, and the latter also provides convenience for the construction of Lyapunov functions. Whereas, considering the proportional delay coupling $\sum_{i=1}^{N} h_{ij}\Gamma x_j(pt)$, which represents that the information interaction occurs at time t and time pt ($0 < p < 1$ is named proportional rate) corresponding to the i-th system state and the j-th system state, respectively. The first case, $p = 0$, which denotes the systems are only coupled in the initial state. The second case, $p = 1$, $\sum_{i=1}^{N} h_{ij}\Gamma x_j(t)$ could be got, which becomes a general state coupling. In the case of $0 < p < 1$, the coupling term could be rewritten as $\sum_{i=1}^{N} h_{ij}\Gamma x_j(t - \tau(t))$ by denoting $pt = t - \tau(t)$ and $\tau(t) = t - pt = (1 - p)t$. Furthermore, Due to $\tau(t) = (1 - p)t \to \infty (t \to \infty)$, pt could be regarded as an unbounded time-varying delay coupling. The problem of synchronization we considered in this paper under the complex dynamical networks with proportional delay. It can be obtained from the above model (4.1) that the current state of the i-th system depends on its own essential state information and the state information of the neighboring system $x_j(t)$ at time pt with proportional rate $p \in (0, 1)$. Obviously, it is more challenging to derive the exponential estimation of (4.19) compared to previous works, such as [18, 20, 21], on this issue. The main reason lies in the time evolution of the complex network is proportionally shifted by the unit pt instead of linearly shifted by normal time unit t. Whereupon a novel impulsive comparison system has been given to coping with this problem. As for the proof process, the formula of parameter variation method has been applied. Meanwhile, we cleverly formulate some parameter equations like (4.23) and (4.27) to prove the contradiction in reference (4.20).

Remark 4.3 For the purpose of making the network model we constructed more truly reflect complexity of the real world and practical applications, the coupling between systems is not only at the current state or historical state, but also at time derivative state. It can be seen in references [35–38] when considering the practical information transmission among different systems in the networks. Meanwhile, spatial position of the i-th system in the network could be represented by the state vectors $x_i(t)$, and the speed of information exchange between systems could be described by $\dot{x}_i(t)$. Furthermore, we can get the dynamics of the i-th system and the velocities of its neighbor systems jointly decide the velocity of the i-th system based on derivative coupled complex dynamical network model (4.1).

Remark 4.4 In fact, the controller (4.4) involves feedback term and impulsive input term. By selecting parameter $\delta \geq 1$ related to impulse effect, we derive adaptively exponentially synchronize complex dynamical network with derivative coupling. Meanwhile, the conclusion that there is a great correlation between parameter δ and impulse effect ρ can be easily obtained according to the $\delta = (1+\rho)^2\lambda_{\max}(I_N - c_1 H)$ defined in Theorem 4.1. In this sense, the relationship between δ and ρ plays an key role in the design of impulsive pinning controller (4.4). In special cases, two inequalities $\rho \geq \sqrt{\frac{1}{\lambda_{\max}(I_N - c_1 H)}} - 1$ or $\rho \leq -\sqrt{\frac{1}{\lambda_{\max}(I_N - c_1 H)}} - 1$ must be satisfied while $\delta \geq 1$. Furthermore, The time-varying control gains $d_i(t)$ will be adaptively updated according to the law (4.5) while the selection of impulsive effects has completed. It can conclude that d_i will adaptively decrease when the impulsive part plays a positive role in the process of network synchronization. However, when the impulse effect destroys the network synchronization, d_i will adaptively increase to reduce this adverse effect.

Remark 4.5 As one of the most popular control strategies, impulsive control strategies have been applied to many engineering fields due to its practicality [16, 17]. During the process of network synchronization, impulsive effect ρ plays a crucial role [8, 13, 14, 21]. Concretely, the works of discussing different function of the impulsive effects respectively, when developing impulsive control protocol have been given in recent years like [18]. Nevertheless, for purpose of speeding up the synchronization of the network, adaptive impulsive control strategy is proposed in this paper. Hence, the size of the the impulsive effect ρ would be limited into a certain range shown in Remark 4.3. In addition, since there exists adaptive related terms $\sum_{i=1}^{N} \frac{1}{2\eta_i}(d_i(t) - d_i)^2$ in the Lyapunov function (4.12), the formula $\delta = (1+\rho)^2\lambda_{\max}(I_N - c_1 H) \geq 1$ becomes indispensable when proceed the inequality (4.13). Furthermore, it is conclude that $\rho \geq \sqrt{\frac{1}{\lambda_{\max}(I_N - c_1 H)}} - 1$ or $\rho \leq -\sqrt{\frac{1}{\lambda_{\max}(I_N - c_1 H)}} - 1$ should be satisfied by impulsive effect ρ for deriving the impulsive equation $y(t_k^+) = \delta y(t_k^-)$ in comparison system (4.15), as well as applying the comparison principle in Lemma 4.1. Otherwise, the control strategy has been proposed cannot be used directly when impulsive effect ρ within $-\sqrt{\frac{1}{\lambda_{\max}(I_N - c_1 H)}} - 1 \leq \rho \leq \sqrt{\frac{1}{\lambda_{\max}(I_N - c_1 H)}} - 1$. Fortunately, control strategies have been proposed in [18] and references therein for this situation.

Remark 4.6 It is easily got from the conditions (4.9) and (4.10) that the time delay in the systems has no effect on whether the network can achieve synchronization. In other words, when analyzing the problem of network synchronization, proportional factor p can be unknown. But when focusing on the exponential convergence rate of the network synchronization, based on $\frac{\lambda}{2}$, the size of p plays a great role. Furthermore, The convergence speed of complex network exponential synchronization is negatively related to the value of P according to the inequalities in (4.11).

4.4 Numerical Simulation

In this section, the correctness of the obtained main results will be verified through giving a numerical simulation without loss of generality

Select the Chua's circuits [42] with following parameters

$$\begin{cases} \dot{s}_1(t) = a(s_2(t) - w(s_1(t))), \\ \dot{s}_2(t) = s_1(t) - s_2(t) + s_3(t), \\ \dot{s}_3(t) = -bs_2(t) - cs_3(t), \end{cases} \tag{4.30}$$

and nonlinear function $w(s_1(t)) = ms_1(t) + \frac{1}{2}(n - m)(|s_1(t) + 1| - |s_1(t) - 1|)$. Then the system parameters chosen as $a = 9.78$, $b = 14.97$, $c = 0$, $n = 1.31$, $m = 0.75$.

In this numerical simulation, the derivative coupled complex network (4.1) with setting the proportional delay rate $p = 0.8$ composed by three Chua's circuits (4.30) is considered. Then $0.8t = pt = t - \tau(t)$, that is, $\tau(t) = (1 - p)t = 0.2t$ can be obtained. In fact, synchronize the three coupled Chua's circuits (4.1) to the isolated Chua's circuits (4.30) under the controller (4.4), which is the objective of this example.

Consider the derivative coupled complex network (4.1) with the following parameters:

$$H = \begin{bmatrix} -1 & 1 & 0 \\ 1 & -2 & 1 \\ 0 & 1 & -1 \end{bmatrix}, \quad B = \begin{bmatrix} -2 & 1 & 1 \\ 1 & -1 & 0 \\ 0 & 1 & -1 \end{bmatrix},$$

the coupling strengths are $c_1 = 0.2$, $c_2 = 0.5$. Suppose that the average impulsive interval N_a is not greater than 0.02, corresponding to the free adjust constant $N_0 = 1$. $\lambda_{\max}(I_N - c_1 H) = 1.6$ can be gotten later. Furthermore, impulsive effect ρ should satisfy $\rho \geq \sqrt{\frac{1}{\lambda_{\max}(I_N - c_1 H)}} - 1$ or $\rho \leq -\sqrt{\frac{1}{\lambda_{\max}(I_N - c_1 H)}} - 1$ based on the above analysis, that is $\rho \geq -0.2094$ or $\rho \leq -1.7906$. So $\rho = 0.8 \geq -0.2094$ are selected. Three suitable feedback control gains are obtained under the adaptive updating laws (4.5). In Fig. 4.2, gives The exact adaptive updating effects of $d_i(t)$.

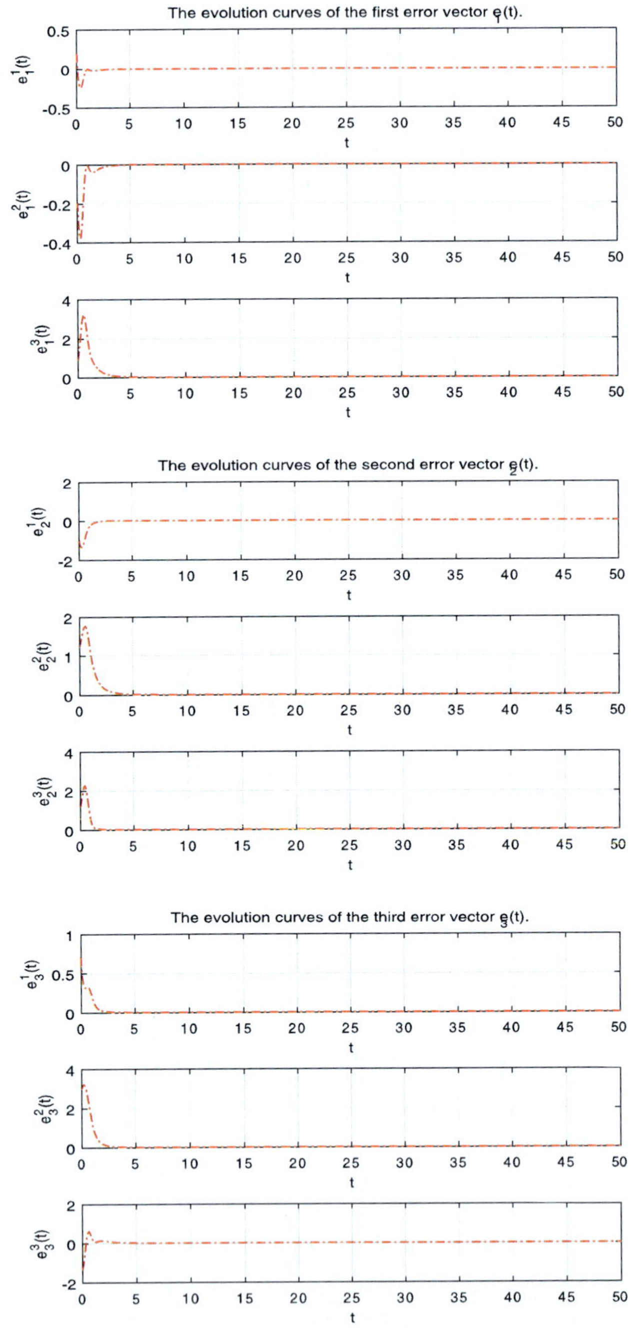

Fig. 4.1 The evolution curves of three different error vectors $e_i^j(t)(i, j = 1, 2, 3)$

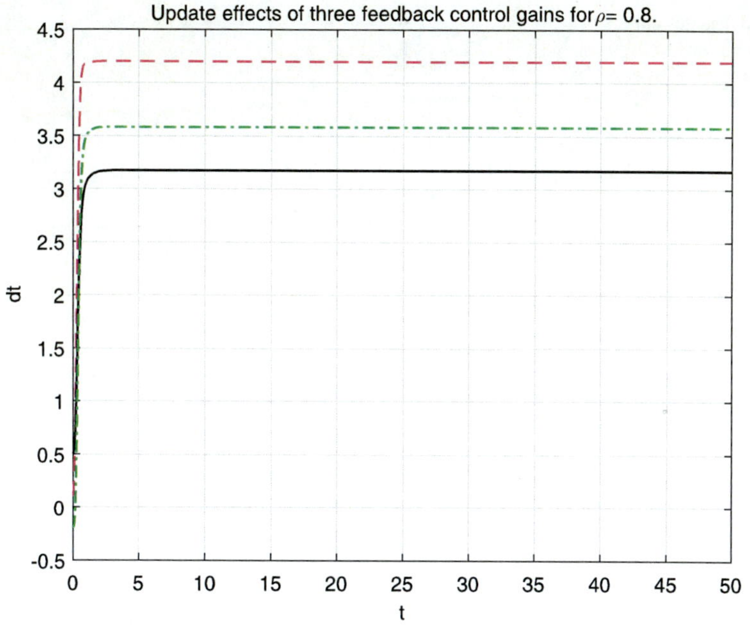

Fig. 4.2 Adaptive updating effects of three feedback control gains $d_i (i = 1, 2, 3)$ when $\rho = 0.8$

The evolution curves of each error vector and error state are shown in Fig. 4.1 according to the definition of error vector

$$x_i(t) - s(t) = e_i(t) = [e_i^1(t), e_i^2(t), e_i^3(t)]^T (i = 1, 2, 3).$$

Moreover, the states curves of the derivative coupled complex networks are presented in Fig. 4.3. Furthermore, define the synchronization error as

$$E(t) = \sqrt{\frac{1}{nN}\sum_{i=1}^{N}\sum_{j=1}^{n}(e_i^j(t))^2}.$$

From the evolution curves of the synchronization error $E(t)$ shown in Fig. 4.4, which can be concluded that the synchronization of the derivative coupled complex networks with proportional delay could be realized within about $8s$.

For the purpose of verifying the effectiveness of the impulsive pinning control strategy (4.4), two more impulsive effects $\rho = 2$ and $\rho = -2$ are be taken, which all satisfy with $\delta \geq 1$. It can be seen in Fig. 4.5, $d_1 = 4, d_2 = 3.4, d_3 = 3.1$ need to be required in case where $\rho = 2$, and for $\rho = -2$, feedback control gains $d_1 = 4.7, d_2 = 3.9, d_3 = 3.5$ are guaranteed.

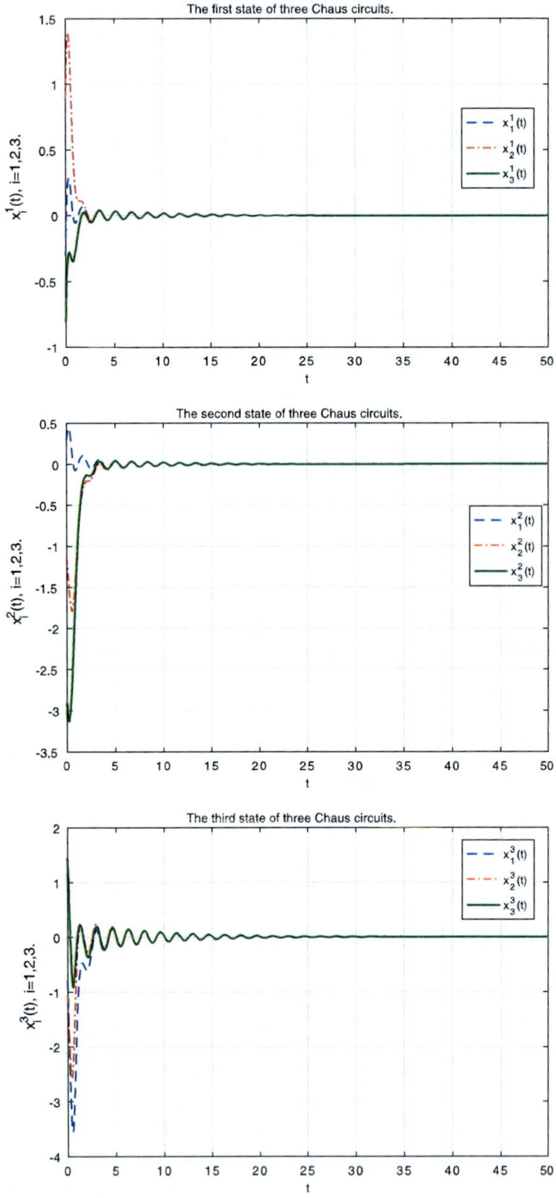

Fig. 4.3 States curves of Chua's circuits in the derivative coupled complex networks with proportional delay

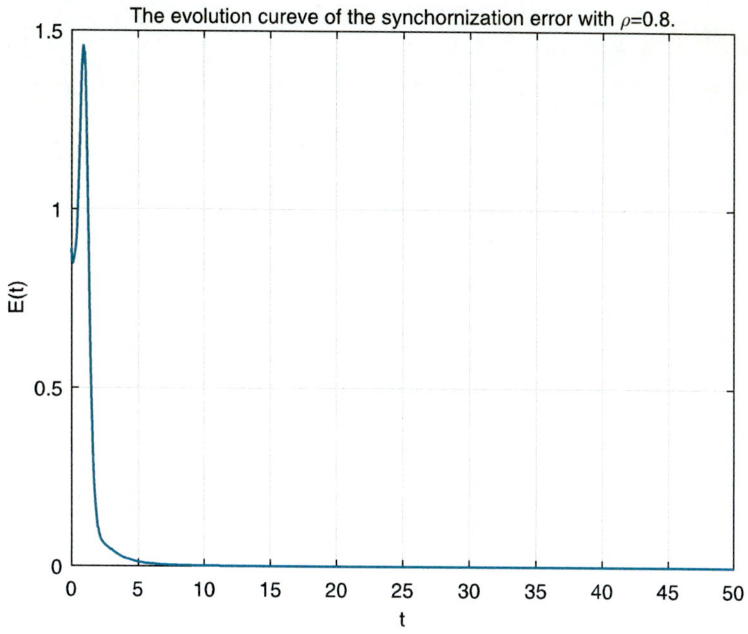

Fig. 4.4 Synchronization error $E(t)$

Remark 4.7 Defining the impulsive distance between the taken impulsive effect ρ and the impulse threshold values when impulsive effects satisfying $\rho \geq -0.2094$ or $\rho \leq -1.7906$. The form as follows

$$d_\rho = \min\{|\rho - \tilde{\rho}|, |\rho - \bar{\rho}|\}.$$

Meanwhile, it can be concluded that the smaller feedback control gain $d_i (i = 1, 2, 3)$ when the larger impulsive distance d_ρ from trying different values of ρ multiple times. In this respect, we get the dynamic balance between the impulsive effects and the feedback control gains. This relationship can provide inspirations for the design of impulsive feedback controller (4.4).

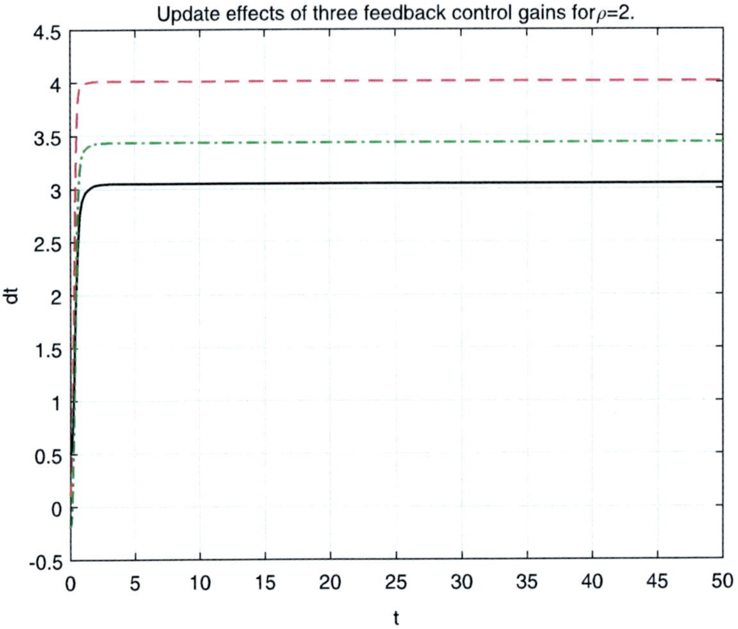

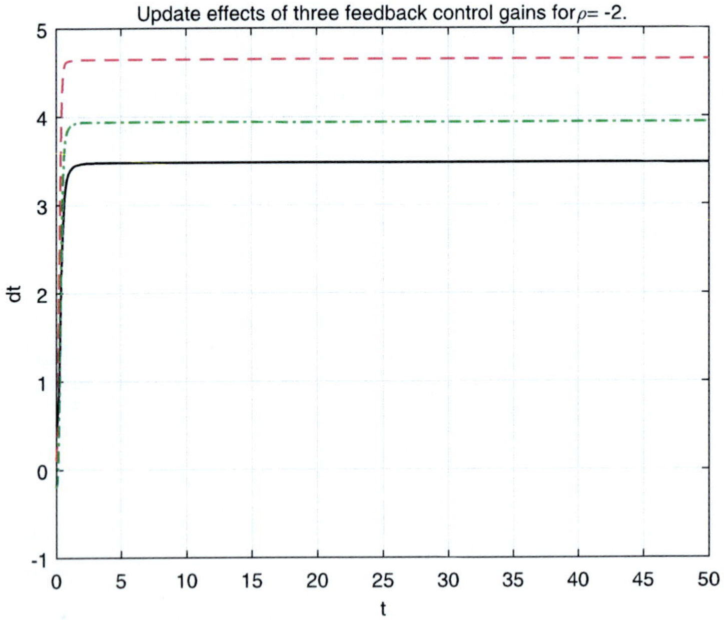

Fig. 4.5 Adaptive updating effects of three feedback control gains $d_i\,(i = 1, 2, 3)$ when $\rho = 2$ and $\rho = -2$, respectively

4.5 Conclusion

The exponential synchronization problem of a class of complex networks with derivative coupling and proportional delay has been discussed in this chapter. Sufficient conditions for exponentially synchronize complex dynamical networks have been obtained via adaptive impulsive pinning control strategy based on the impulsive comparison principle, the extended parameters variation formula and the concept of average impulsive interval. Furthermore, according to the efficiently designed adaptive updating laws, suitable feedback control gains and exponential convergence rate have been gotten. Finally, the effectiveness of the main results has been verified by presenting a numerical example. Particularly, other uncertainties like stochastic perturbations, time delays would be considered in our future works.

References

1. Strogatz, S. H.: Exploring complex networks. Nature. **410**, 268–276 (2001)
2. Li, C. G., Chen, G. R.: Synchronization in general complex dynamical networks with coupling delays. Physica A: Statistical Mechanics and its Applications. **343**, 263–278 (2004)
3. Zhang, D., Shi, P., Yu, L.: Containment control of linear multiagent systems with aperiodic sampling and measurement size reduction. IEEE Transactions on Neural Networks and Learning Systems. **29**, 5020–5029 (2018)
4. Qi, W. H., Zong, G. D., Karimi, H. R.: Observer-based adaptive SMC for nonlinear uncertain singular semi-Markov jump systems with applications to DC motor. IEEE Transactions on Circuits and Systems I: Regular Papers. **65**, 2951–2960 (2018)
5. Zhang, D., Xu, Z. H., Srinivasan, D., Yu, L.: Leader-follower consensus of multiagent systems with energy constraints: A Markovian system approach. IEEE Transactions on Systems, Man, and Cybernetics: System. **47**, 1727–1736 (2017)
6. Qi, W. H., Park, J. H., Zong, G. D., Cao, J. D., Cheng, J.: A fuzzy Lyapunov function approach to positive L1 observer design for positive fuzzy semi-Markovian switching systems with its application. IEEE Transactions on Systems, Man, and Cybernetics: Systems. **51**, 775–785 (2021)
7. Song, X. M., Yan, X. H.: Linear quadratic Gaussian control for linear time-delay systems. IET Control Theory and Applications. **8**, 375–383 (2014)
8. He, W. L., Chen, G. R., Han, Q. L., Qian, F.: Network-based leader-following consensus of nonlinear multi-agent systems via distributed impulsive control. Information Science. **380**, 145–158 (2017)
9. Guan, Z. H., Liu, Z. W., Feng, G., Wang, Y. W.: Synchronization of complex dynamical networks with time-varying delays via impulsive distributed control. IEEE Transactions on Circuits Systems I: Regular Papers. **57**, 2182–2195 (2010)
10. Chen, Y., Yu, W. W., Li, F. F., Feng, S. S.: Synchronization of complex networks with impulsive control and disconnected topology. IEEE Transactions on Circuits Systems II: Express Briefs. **60**, 292–296 (2013)
11. Xu, W. Y., Ho, D. W. C.: Clustered event-triggered consensus analysis: An impulsive framework. IEEE Transactions on Industrial Electronics. **63**, 7133–7143 (2016)
12. Chen, W. H., Jiang, Z. Y., Lu, X. M., Luo, S. X.: H_∞ synchronization for complex dynamical networks with coupling delays using distributed impulsive control. Nonlinear Analysis: Hybrid Systems. **17**, 111–127 (2015)

13. Lu, J. Q., Ho, D. W. C., Cao, J. D., Kurths, J.: Single impulsive controller for globally exponential synchronization of dynamical networks. Nonlinear Analysis: Real World Applications. **14**, 581–593 (2013)
14. Yang, Z. C., Xu, D. Y.: Stability analysis and design of impulsive control systems with time delay. IEEE Transactions on Automatic Control. **52**, 1448–1454 (2007)
15. Hu, J. Q., Liang, J. L., Cao, J. D.: Synchronization of hybrid-coupled heterogeneous networks: pinning control and impulsive control schemes. Journal of Franklin Institute. **351**, 2600–2622 (2014)
16. Yang, T., Chua, L. O.: Impulsive stabilization for control and synchronization of chaotic systems: theory and application to secure communication. IEEE Transactions on Circuits and Systems I: Fundamental Theory and Applications. **44**, 976–988 (1997)
17. Yang, T.: Impulsive control. IEEE Transactions on Automation Control. **44**, 1081–1083 (1999)
18. Tang, Z., Park, J. H., Feng, J. W.: Impulsive effects on quasi-synchronization of neural networks with parameter mismatches and time-varying delay. IEEE Transactions on Neural Networks and Learning Systems. **29**, 908–919 (2018)
19. Yan, X. H., Song, X. M., Wang, X. H.: Global output-feedback stabilization for nonlinear time-delay systems with unknown control coefficients. International Journal of Control Automation and Systems. **16**, 1550–1557 (2018)
20. He, W. L., Qian, F., Han, Q. L., Cao, J. D.: Lag quasi-synchronization of coupled delayed systems with parameter mismatch. IEEE Transactions on Circuits Systems I: Regular Papers. **58**, 1345–1357 (2011)
21. He, W. L., Qian, F., Lam, J., Chen, G. R., Han, Q. L., Kurths, J.: Quasi-synchronization of heterogeneous dynamic networks via distributed impulsive control: Error estimation, optimization and design. Automatica. **62**, 249–262 (2015)
22. Mohajerpoor, R., Shanmugam, L., Abdi, H., Rakkiyappan, R., Nahavandi, S. P. Shi.: New delay range-dependent stability criteria for interval time-varying delay systems via Wirtinger-based inequalities. International Journal of Robust and Nonlinear Control. **28**, 661–677 (2018)
23. Mohajerpoor, R., Shanmugam, L., Abdi, H., Rakkiyappan, R., Nahavandi, S., Park, J. H.: Improved delay-dependent stability criteria for neutral systems with mixed interval time-varying delays and nonlinear disturbances. Journal of the Franklin Institute. **354**, 1169–1194 (2017)
24. Wu, Z. G., Shi, P., Su, H. Y., Chu, J.: Stochastic synchronization of Markovian jump neural networks with time-varying delay using sampled data. IEEE Transactions on Cybernetics. **43**, 1796–1806 (2013)
25. Zhong, J., Lu, J. Q., Liu, Y., Cao, J. D.: Synchronization in an array of output-coupled Boolean networks with time delay. IEEE Transactions on Neural Networks and Learning Systems. **25**, 2288–2294 (2014)
26. Lu, J. Q., Zhong, J., Ho, Daniel W. C., Tang, Y., Cao, J. D.: On controllability of delayed Boolean control networks. SIAM Journal of Control Optimization. **54**, 475–494 (2016)
27. Wang, L. M., Shen, Y., Zhang, G. D.: Synchronization of a class of switched neural networks with time-varying delays via nonlinear feedback control. IEEE Transactions on Cybernetics. **46**, 2300–2310 (2016)
28. Yang, X. S., Ho, Daniel W. C.: Synchronization of delayed memrisitive neural networks: robust analysis approach. IEEE Transactions on Cybernetics. **46**, 3377–3387 (2016)
29. Zhou, L. Q., Zhang, Y. Y.: Global exponential stability of cellular neural networks with multi-proportional delays. International Journal of Biomathematics. **8**, 1–17 (2015)
30. Zhou, W. N., Wang, T. B., Zhong, Q. C., Fang, J. A.: Proportional-delay adaptive control for global synchronization of complex networks with time-delay and switching outer-coupling matrices. International Journal of Robust and Nonlinear Control. **23**, 548–561 (2013)
31. Zheng, C., Li, N., Cao, J. D.: Matrix measure based stability criteria for high-order neural networks with proportional delay. Neurocomputing. **149**, 1149–1154 (2015)
32. Jia, R. W.: Finite-time stability of a class of fuzzy cellular neural networks with multi-proportional delays. Fuzzy Sets and Systems. **319**, 70–80 (2017)

33. Zhou, L. Q.: Delay-dependent exponential synchronization of recurrent neural networks with multiple proportional delays. Neural Process Letters. **42**, 619–632 (2015)
34. Zhou, L. Q., Chen, X. B., Yang, Y. X.: Asymptotic stability of cellular neural networks with multiple proportional delays. Applied Mathematical Computation. **229**, 457–466 (2014)
35. Xu, Y. H., Zhou, W. N., Fang, J. A., Xie, C. R., Tong, D. B.: Finite-time synchronization of the complex dynamical network with non-derivative and derivative coupling. Neurocomputing. **173**, 1356–1361 (2016)
36. Zheng, S.: Pinning and impulsive synchronization control of complex dynamical networks with non-derivative and derivative coupling. Journal of the Franklin Institute. **354**, 6341–6363 (2017)
37. Xu, Y. H., Zhou, W. N., Fang, J. A.: Sun, W., Pan, L.: Topology identification and adaptive synchronization of uncertain complex networks with non-derivative and derivative coupling. Journal of the Franklin Institute. **347**, 1566–1576 (2010)
38. Deng, L. P., Wu, Z. Y., Wu, Q. C.: Pinning synchronization of complex network with non-derivative and derivative coupling. Nonlinear Dynamics. **73**, 775–782 (2013)
39. Bian, Q. X., Yao, H. X.: Adaptive synchronization of bipartite dynamical networks with distributed delays and nonlinear derivative coupling. Communications in Nonlinear Science and Numerical Simulation. **16**, 4089–4098 (2011)
40. Xu, Y. H., Zhou, W. N., Fang, J. A., Sun, W.: Adaptive synchronization of the complex dynamical network with non-derivative and derivative coupling. Physics Letters A. **374**, 1673–1677 (2010)
41. Lakshmikantham, V., Bainov, D., Simenonv, P.: Theory of impulsive different equations. World Scientific, Singapore (1989)
42. Chua, L. O., Itoh, M., Kocarev, L., Eckert, K.: Chaos synchronization in Chua's circuit. Journal of Circuits, Systems and Computers. **3**, 93–108 (1993)

Chapter 5
Distributed Impulsive Quasi-Synchronization of Lur'e DNs with Proportional Delay

5.1 Introduction

In the past few years, complex networks have been studied by many researchers from many kinds of fields [1–3]. The collective behaviors of a complex dynamical network became a hot topic for its practical merits in engineering fields during recent years. Among these collective behaviors of complex networks, the investigation on synchronization, which represents the special uniform behavior of the systems in the networks, becomes more and more important. As a meaningful part of practical world, the synchronization has many applications like transportation network, image processing and so on.

In fact, many complex networks could not realize synchronization through changing their own parameters. In order to settle this matter, interactions of internal subsystems and suitable parameters are usually required in this situation such as [4, 5]. On the contrary, under the outside control, the synchronization of complex networks could also be achieved [6–8].

If there exist some unavoidable disturbances in the network or the decreased requirements on synchronization, such as mismatched components, failures of communication and the outside attacks, the synchronization error could only converges into a finite bound. This kind of bounded synchronization phenomenons is called quasi-synchronization which implies the synchronization error is always under certain error upper bound [9–17], whether it is controlled or not. For example, in [11], Liu et al. studied the bound synchronization of delayed neural networks with discontinuous activation functions. In addition, He et al. studied the synchronization of complex networks under distributed impulsive control within a nonzero error bound in [17].

Especially, in some practical situations, the communication channels often suffer from some disturbances when the information transmissions and exchanges among different systems happen because of some unavoidable uncertainties, such as the external environmental change, interruption of communication, the limited data

Z. Tang et al., *Impulsive Synchronization of Complex Dynamical Networks*,
https://doi.org/10.1007/978-981-16-5383-4_5

capacity and so on. As an important factor, time delay heavily affects the behaviors of the corresponding systems and fully taking into account by many scholars when the mathematical model of complex networks with the all kinds of time delays is established [18–22]. Meanwhile, it is assumed that most of the considered delays satisfy some strict conditions, like delays with prescribed bound $0 \leq \tau(t) \leq \tau$. Different from the general delay, proportional delay is usually viewed as a kind of unbounded delays and utilized in some practical situation [23–25]. Complex dynamical networks with proportional delays are usually considered as the quality of service models for its unique characteristics like controllable and predictable. Generally, a practical mathematical model of the complex network always contains the proportional delay.

Particularly, the current states and historical states of the i-th system like $z_i(t)$ and $z_i(qt)$, are usually denoted as its dynamical behaviors at a point in time, where the time ratio between two states is $q \in (0, 1)$. One can conclude that the historical state of the i-th system could be viewed as the proportion of current state from the definition of proportional delay. Contrast to the time invariant delay τ or time-varying delay $\tau(t)$, the proportional delay is a kind of unbounded delays with the proportional factor q. Therefore, comparing with the previous works like [10], it is more challengeable to synchronize complex networks with proportional delay. Note that the time evolution transform proportionally with the time ratio q instead of general time-varying delay $t - \tau(t)$. To settle proportional delay issue on the synchronization of networks, works [26–28] showed some previous results. For example, Zhou et al. studied the exponential synchronization of the cellular neural networks with proportional delays via introducing the Brouwer fixed point theorem in [23]. Meanwhile, Jia et al. investigated the synchronization problem of fuzzy cellular neural networks with multiple proportional delays in [26].

Generally speaking, as a discontinuous control strategy, the impulsive control provides a kind of instant motion power to the synchronization process of complex networks and saves energy compared with general continuous control [29, 30]. For the impulsive issues, impulses could either be viewed as the outside attacks or the positive control to the corresponding networks. In impulsive synchronization process, impulsive effect is undoubtedly a influential part like [31]. So far, the influence of impulsive effects has been studied in many existing works on the impulsive synchronization problems [32–35]. For instance, scholars discussed the network-based leader-following consensus of multi-agent systems via introducing a distributed impulsive controller in [32]. Lu et al. in [33] studied the stabilization issue of the complex networks with time-varying delay under the pinning control. Zhang et al. in [34] studied the synchronization of complex networks with stochastic factors under the delayed impulsive controller. Specially, it is also necessary to consider negative effect caused by some impulsive effects in the network.

It's worth noting that the impulsive synchronization problem of complex dynamical networks with proportional delays seldom receives attention from scholars despite its significance until now. And motivated by the complication in theoretical proof and the importance in practical applications, current work is proposed to solve this problem. In this chapter, the quasi-synchronization of a kind of coupled complex networks is investigated via introducing a distributed impulsive pinning controller.

The distinct merits of this chapter can be presented as follow: (1) Different from models studied in previous works with all kinds of time delays, the mathematical model of coupled Lur'e network with the proportional delay is established, where proportional delay is a kind of unbounded delays. (2) Fully taking the existence of mismatched system parameters into account, criteria for the bounded synchronization of complex dynamical networks consisted of nonlinear Lur'e systems is derived according to the impulsive comparison principle, the extended formula for the variation of parameters and the concept of average impulsive interval in this chapter. (3) In order to achieve quasi-synchronization of the Lur'e networks, a kind of distributed impulsive pinning controllers has been designed. (4) Considering the positive and negative effect caused by impulses, the synchronization errors and the convergence rates are calculated via some given parameter functions. Additionally, three numerical simulations are proposed to show the validity of control protocol and theoretical results.

The rest sections are arranged as follows. In Sect. 5.2, some necessary preliminaries and the mathematical model of the networks are given. In Sect. 5.3, the investigation on the exponential quasi-synchronization of Lur'e networks is expressed. In Sect. 5.4, Several examples are shown to prove the validity of main results in this chapter. In the end, the conclusion of this chapter are expressed in Sect. 5.5.

Notations. $\mathbb{R}^n$ denotes the n-dimensional Euclidean space. $\mathbb{R}^{n\times n}$ is the set of $n \times n$ real matrices. diag$\{\cdots\}$ presents a diagonal matrix. $\|\cdot\|$ stands for the Euclid norm of the matrix or the vector. $\lambda_{\max}(H)$ stands for the largest eigenvalue of matrix H and $\max_{1\leq i\leq N}\{\cdot\}$ denotes the maximum value. $H > 0 (H \geq 0)$ denotes the positive definite (semi-definite) matrix H. I_n stands for the identity matrix with n dimension. $H = (h_{ij})$ denotes a $\mathbb{R}^{N\times N}$ matrix H with elements h_{ij} for $i, j = 1, 2, \ldots, N$. The set $\mathcal{N}$ denotes the positive integers. The dimension of these vectors and matrices will be cleared in the context.

5.2 Model Description and Preliminaries

5.2.1 Model Description

Consider the following complex dynamical network consisting of nonidentical Lur'e systems and proportional delay

$$\dot{z}_i(t) = A_i z_i(t) + B_i \tilde{f}(Cz_i(t)) + c\sum_{j=1}^{N} g_{ij}\Gamma z_j(qt), \tag{5.1}$$

where the state vector is denoted as $z_i(t) = [z_i^1(t), z_i^2(t), \ldots, z_i^n(t)]^T \in \mathbb{R}^n$ for $i = 1, 2, \ldots, N$ in the i-th Lur'e system. $A_i \in \mathbb{R}^{n\times n}$ and $B_i \in \mathbb{R}^{n\times m}$ are constant matrices. Matrix $C \in \mathbb{R}^{m\times n}$ is defined as $C = [c_1, c_2, \ldots, c_m]^T$ with $c_j \in \mathbb{R}^{1\times n}$ for

$j = 1, 2, \ldots, m$. And then $Cz_i(t) = [c_1 z_i(t), c_2 z_i(t), \ldots, c_m z_i(t)]^T$, $\tilde{f}(Cz_i(t)) = [\tilde{f}_1(c_1 z_i(t)), \tilde{f}_2(c_2 z_i(t)), \ldots, \tilde{f}_m(c_m z_i(t))]^T$. The constant $c > 0$ represents the coupling strength. $\Gamma = \text{diag}\{r_1, r_2, \ldots, r_n\} \in \mathbb{R}^{n\times n}$ is the inner connection matrix with $r_i \geq 0$. Without loss of generality, we assume that $\Gamma = I_n$ in this chapter. Function $\tilde{f} : \mathbb{R}^m \to \mathbb{R}^m$ is the memoryless nonlinear vector-valued function which is continuous differentiable on $\mathbb{R}$. Matrix $G = (g_{ij}) \in \mathbb{R}^{N\times N}$ is the outer connection matrix which is subject to the topology structure, and it is assumed to satisfy diffusive condition $g_{ii} = -\sum_{j\neq ij=1}^{N} g_{ij} = -\sum_{j\neq ij=1}^{N} g_{ji}$, where $g_{ji} = g_{ij} > 0$ if there is a connection between the i-th Lur'e system and the j-th Lur'e system for $i \neq j$, and $g_{ij} = 0$ otherwise. The factor $q \in (0, 1)$ is a constant which involves the history time. Especially, dynamics of the i-th Lur'e system is decided by the state vectors $z_j(t)$ for $j = 1, 2, \ldots, N$, and the states at history time $z_j(qt)$ could be regard as the proportional state from the current state vector with a time ratio q in the Lur'e network model (5.1). Thus, q is viewed as proportional delay and we could get that $qt = t - \tau(t)$ for $\tau(t) = (1-q)t \geq 0$ but $\tau(t) \to \infty$. It follows that the proportional delay could be regarded as a kind of time-varying delay without bound.

Consider the following isolated Lur'e system with different system parameters from (5.1)

$$\dot{x}(t) = Ax(t) + B\tilde{f}(Cx(t)), \tag{5.2}$$

where $x(t) = [x^1(t), x^2(t), \ldots, x^n(t)]^T \in \mathbb{R}^n$. Besides, the solution $x(t)$ of the Lur'e system (5.2) and other Lur'e systems in network (5.1) could be viewed as the leader and its followers, respectively. Thus, the synchronization between the Lur'e dynamical network (5.1) and the Lur'e system (5.2) can be thought as a class of leader-following issue.

In this chapter, for a prescribed positive constant Q and a time instant T_0, there holds $\|x(t)\| \leq Q$ for all $t \geq T_0$ and any initial $x(0)$, that is $x(t)$ is bounded with boundary Q.

Define the error vector as $\Delta z_i(t) = z_i(t) - x(t) = [\Delta z_i^1(t), \Delta z_i^2(t), \ldots, \Delta z_i^n(t)]^T \in \mathbb{R}^n$ for $i = 1, 2, \ldots, N$. From the Lur'e dynamical network (5.1) and the Lur'e system (5.2), the coupled error Lur'e network with outside controller $u_i(t)$ could be described as

$$\begin{aligned}\Delta\dot{z}_i(t) = {} & A_i\Delta z_i(t) + B_i f(C\Delta z_i(t)) + c\sum_{j=1}^{N} g_{ij}\Delta z_j(qt) \\ & + u_i(t) + Y_i(z(t)),\ i = 1, 2, \ldots, N,\end{aligned} \tag{5.3}$$

where function $f(C\Delta z_i(t)) = \tilde{f}(Cz_i(t)) - \tilde{f}(Cz(t))$ and parameter mismatch $Y_i(z(t)) \in \mathbb{R}^n$ which is so-called heterogeneities between Lur'e network (5.1) and (5.2), is defined as $Y_i(z(t)) = (A_i - A)z(t) + (B_i - B)\tilde{f}(Cz(t))$.

Generally, for the purpose of simulating the outside disturbance, the controller $u_i(t)$ with the coefficient μ is designed in this chapter. Therefore, the controlled error Lur'e network could be rewritten as

$$\begin{aligned}\dot{\Delta} z_i(t) &= A_i \Delta z_i(t) + B_i f(C \Delta z_i(t)) + c \sum_{j=1}^{N} g_{ij} \Delta z_j(qt) \\ &\quad + u_i(t) + Y_i(z(t)) + \mu \sum_{k=1}^{+\infty} u_i(t) \delta(t - t_k), \end{aligned} \tag{5.4}$$

where μ is impulsive effect and $\delta(\cdot)$ is the Dirac function. The impulsive sequence $\zeta = \{t_1, t_2, \ldots\}$ is a increasing series of impulsive times which satisfies $t_{k-1} < t_k$ with $\lim_{k \to +\infty} t_k = +\infty$ for $k \in \mathbb{N}$.

Considering the distributed strategy, for the purpose of quasi-synchronization between the Lur'e network (5.1) and (5.2), the pinning controller is thus elaborately designed as

$$u_i(t) = k \sum_{j \in \mathbb{N}_i} w_{ij}(z_j(t) - z_i(t)) - d_i \Delta z_i(t), \tag{5.5}$$

where $\mathbb{N}_i$ contains other Lur'e systems which link to the i-th system. $W = (w_{ij}) \in \mathbb{R}^{N \times N}$ is the coupling matrix satisfying the diffusive condition, that is $w_{ii} = -\sum_{j \neq i j=1}^{N} w_{ij} = -\sum_{j \neq i j=1}^{N} w_{ji}$, where $w_{ji} = w_{ij} > 0$ if there is a connection between the i-th Lur'e system and the j-th Lur'e system $(i \neq j)$ and $w_{ij} = 0$, otherwise. Nonnegative parameters k and d_i $(i = 1, 2, \ldots, N)$ are control gains, besides, at least one $d_i > 0$. Moreover, the feedback control gain matrix is denoted as $D = \text{diag}\{d_1, d_2, \ldots, d_N\}$.

According to the above discussion, the following impulsive error Lur'e dynamical network with proportional delay can be derived as

$$\begin{cases} \dot{\Delta} z_i(t) = A_i \Delta z_i(t) + B_i f(C \Delta z_i(t)) + Y_i(z(t)) \\ \qquad + c \sum_{j=1}^{N} g_{ij} \Delta z_j(qt) - d_i \Delta z_i(t) \\ \qquad + k \sum_{j=1}^{N} w_{ij} \Delta z_j(t), t \neq t_k, k \in \mathbb{N}, \\ \Delta z_i(t_k) = \mu(k \sum_{j=1}^{N} w_{ij} \Delta z_j(t_k^-) - d_i \Delta z_i(t_k^-)), k \in \mathbb{N} \\ \Delta z_i(0) = \psi_i(0), i = 1, 2, \ldots, N, \end{cases} \tag{5.6}$$

where the error state vector $\Delta z_i(t)$ is assumed to be right-hand continuous at $t = t_k, k \in \mathbb{N}$, and $\Delta z_i(t_k) = \Delta z_i(t_k^+) = \lim_{\epsilon \to 0^+} \Delta z_i(t_k + \epsilon)$, $\Delta z_i(t_k^-) = \lim_{\epsilon \to 0^-} \Delta z_i(t_k + \epsilon)$ throughout this chapter. Thus, the solutions of (5.6) are piecewise right-hand continuous functions with discontinuities at impulsive time $t = t_k$ for $k \in \mathbb{N}$.

5.2.2 Preliminaries

In order to derive the main results of this chapter, several necessary definitions, lemmas and assumptions are expressed in this section.

Definition 5.1 ([15]) Taking the coupled Lur'e networks (5.1) and target Lur'e system (5.2) into account. If there exists a compact set $\bar{C}$ such that the error vector $\Delta z_i(t)$ converges into the ball $\bar{C} = \{\Delta z(t) = [\Delta z_1^T(t), \Delta z_2^T(t), \ldots, \Delta z_N^T(t)]^T \in \mathbb{R}^{nN} | \|\Delta z(t)\| \leq \bar{z}\}$ as $t \to +\infty$ for any initial conditions ψ_i^0, then the synchronization of the nonlinear error Lur'e network (5.6) is realized within a prescribed error bound $\bar{z} > 0$.

Definition 5.2 ([33]) Denote a impulsive sequence as $\zeta = \{t_1, t_2, \ldots\}$ and the number of impulses as $N_\zeta(T, t)$ in the time interval (t, T). If there exist two positive scalars N_0 and N_a, such that the following inequality satisfies

$$\frac{T-t}{N_a} - N_0 \leq N_\zeta(T, t) \leq \frac{T-t}{N_a} + N_0, \forall\, T \geq t \geq 0, \tag{5.7}$$

where the average impulsive interval is less than N_a for the impulsive sequence ζ.

Lemma 5.1 ([35]) *Denote $\varphi(t)$ as a continuous function with some finite discontinuity points t_k at which $\varphi(t_k^+) = \varphi(t_k)$ and $\varphi(t_k^-)$ exist. Let the set $PC(l)$ is $\{\varphi | \varphi : [-\tau, \infty) \to \mathbb{R}^l\}$ and $0 \leq \tau(t) \leq \tau$. Therefore, $u(t)$ and $v(t)$ are chosen from the above set $PC(l)$ with $l = 1$. If there exist three constants a, b and ρ, such that*

$$\begin{cases} D^+u(t) \leq au(t) + bu(t - \tau(t)),\ t \neq t_k, k \in \mathbb{N}, \\ u(t_k) \leq \rho u(t_k^-),\ k \in \mathbb{N}, \end{cases}$$

and

$$\begin{cases} D^+v(t) > av(t) + bv(t - \tau(t)),\ t \neq t_k, k \in \mathbb{N}, \\ v(t_k) = \rho v(t_k^-),\ k \in \mathbb{N}. \end{cases}$$

Above all, if there exists $u(t) \leq v(t)$ for $t \in [-\tau, 0]$, then there obtains $u(t) \leq v(t)$ for $t \in (0, +\infty)$.

Assumption 5.1 Suppose that the nonlinear function $\tilde{f}(\cdot)$ satisfies the Lipschitz conditions, i.e., there exists a positive constant h, such that

$$\|\tilde{f}(x) - \tilde{f}(y)\| \leq h\|x - y\|$$

for any $x, y \in \mathbb{R}^n$.

Remark 5.1 It is noted that the nonlinear function $Y_i(z(t))$ presents the heterogeneity which caused by mismatched parameters between the nonidentically coupled Lur'e networks (5.1) and the target Lur'e system (5.2) in error Lur'e network (5.3).

Considering the bounded initial condition of $z(t)$ and nonlinear dynamics function $\tilde{f}(\cdot)$ satisfying Lipschitz condition, therefore, the bounded function $Y_i(z(t))$ yields that $\sup_{t\geq\bar{T}} \|Y_i(z(t))\| = y_i$ where nonnegative constants y_i and $\bar{T} > T_0$ for $i = 1, 2, \ldots, N$.

5.3 Main Results

The sufficient criteria for the realization of the exponential quasi-synchronization between the Lur'e networks (5.1) and the target Lur'e system (5.2) will be derived with the comparison principle, the formula for variation of parameters under the distributed pinning controller (5.5).

Theorem 5.1 *From Assumption 5.1, taking the impulsive error Lur'e networks (5.6) into account. In view of the impulsive sequence* $\zeta = \{t_1, t_2, \ldots\}$ *defined in Definition 5.2, suppose that the bound of average impulsive interval is less than* N_a. *If there exist matrices* $D > 0$, $W > 0$ *and the positive scalars* ω, α, β, l_i *such that*

Case I*: For the positive constant* $\omega \leq 1$, *if matrix meets*

(i)
$$\begin{bmatrix} -\omega I_{Nn} & \mu(kW - D)\otimes I_n + I_{Nn} \\ * & -I_{Nn} \end{bmatrix} < 0; \tag{5.8}$$

(ii)
$$\Pi = \begin{bmatrix} kW + aI_N - D & \frac{1}{2}cG & \frac{1}{2}I_N \\ * & -bI_N & 0 \\ * & * & -L \end{bmatrix} < 0; \tag{5.9}$$

(iii-a)
$$\begin{cases} \xi + \lambda q + \beta\omega^{-N_0} < 0, \\ \lambda(\xi + \lambda q + q\beta\omega^{-N_0}) + \xi(\xi + \lambda q + \beta\omega^{-N_0}) > 0, \\ \lambda^2(\xi + \lambda q + q\beta\omega^{-N_0}) > \xi^2(\xi + \lambda q + \beta\omega^{-N_0}) \end{cases} \tag{5.10}$$

for a positive constant $\lambda \in (0, -\xi)$, *then the trajectory of the controlled nonidentical Lur'e networks (5.6) exponentially converges into the compact set* $\bar{C}$ *with the convergence rate* $\frac{\lambda}{2}$, *where the compact set* $\bar{C}$ *could be described as*

$$\bar{C} = \left\{ \Delta z(t) \| \|\Delta z(t)\| \leq \bar{z} = \sqrt{\frac{2\omega^{-N_0}\sum_{i=1}^{N} l_i y_i^2}{-(\xi + \beta\omega^{-N_0})}} \right\}, \tag{5.11}$$

where $\alpha = -\max_{1\leq i\leq N}\{\lambda_{\max}(A_i + A_i^T + B_i^T B_i + h^2 C^T C - 2aI_n)\}$, $\beta = 2b$, $\xi = \frac{\ln\omega}{N_a} - \alpha < 0$ *and* $L = diag\{l_1, l_2, \ldots, l_N\}$. *That is, the quasi-synchronization between the Lur'e networks (5.1) and the target Lur'e system (5.2) is realized with the error bound* $\bar{z}$ *under the distributed pinning controller (5.5).*

***Case II**: Considering conditions **(i)** and **(ii)** in **Case I** are satisfied and*

$$\textbf{\textit{(iii-b)}}\qquad \begin{cases} \xi + \bar{\lambda} q + \beta\omega^{N_0} < 0, \\ \bar{\lambda}((\xi + \bar{\lambda} q) + q\beta\omega^{N_0}) + \xi((\xi + \bar{\lambda} q) + \beta\omega^{N_0}) > 0, \\ \bar{\lambda}^2((\xi + \bar{\lambda} q) + q\beta\omega^{N_0}) > \xi^2((\xi + \bar{\lambda} q) + \beta\omega^{N_0}) \end{cases} \tag{5.12}$$

for $\omega > 1$ and a positive constant $\bar{\lambda} \in (0, -\xi)$, then the solution of the error Lur'e networks (5.6) exponentially converge into the compact set $\tilde{C}$ at the convergence rate $\frac{\bar{\lambda}}{2}$, where the compact set $\tilde{C}$ could be presented as

$$\tilde{C} = \left\{ \Delta z(t) | \|\Delta z(t)\| \leq \tilde{z} = \sqrt{\frac{2\omega^{N_0} \sum_{i=1}^{N} l_i y_i^2}{-(\xi + \beta\omega^{N_0})}} \right\}. \tag{5.13}$$

Then, the quasi-synchronization between the nonidentical Lur'e networks (5.1) and the target Lur'e system (5.2) is achieved within the error range $\tilde{z}$ under the distributed pinning controller (5.5).

Proof Construct the following Lyapunov function

$$V(t) = \frac{1}{2} \sum_{i=1}^{N} \Delta z_i^T(t) \Delta z_i(t), \tag{5.14}$$

and denote $\Delta z(t) = [\Delta z_1^T(t), \Delta z_2^T(t), \ldots, \Delta z_N^T(t)]^T \in \mathbb{R}^{Nn}$ for convenience.

Firstly, based on the impulsively controlled error Lur'e network (5.6) for $t = t_k, k \in \mathbb{N}$, the following results can be derived as

$$\Delta z_i(t_k^+) = \mu k \sum_{j=1}^{N} w_{ij} \Delta z_j(t_k^-) - \mu d_i \Delta z_i(t_k^-) + \Delta z_i(t_k^-).$$

Considering the above equation and Kronecker product, it obtains following result

$$\Delta z(t_k^+) = (\mu(kW - D) \otimes I_n + I_{Nn}) \Delta z(t_k^-).$$

Therefore, function (5.14) can be rewritten as

$$\begin{aligned} V(t_k^+) &= \frac{1}{2} \Delta z^T(t_k^+) \Delta z(t_k^+) \\ &= \frac{1}{2} \Delta z(t_k^-)^T (\mu(kW - D) \otimes I_n + I_{Nn})^T \\ &\quad \times (\mu(kW - D) \otimes I_n + I_{Nn}) \Delta z(t_k^-). \end{aligned}$$

In terms of the linear matrix inequality in condition (5.8) of Theorem 5.1, we have

$$(\mu(kW - D) \otimes I_n + I_{Nn})^T(\mu(kW - D) \otimes I_n + I_{Nn}) < \omega I_{Nn}.$$

It further derives

$$V(t_k^+) < \omega \cdot V(t_k^-). \tag{5.15}$$

Secondly, for $t \in [t_{k-1}, t_k)$ with $k \in \mathbb{N}$, the derivative of (5.14) along the error Lur'e network (5.6) can be calculated as

$$\begin{aligned}
\dot{V}(t) &= \sum_{i=1}^{N} \Delta z_i^T(t)(A_i\Delta z_i(t) + B_i f(C\Delta z_i(t)) + Y_i(z(t)) \\
&\quad + c\sum_{j=1}^{N} g_{ij}\Delta z_j(qt) + k\sum_{j=1}^{N} w_{ij}\Delta z_j(t) - d_i\Delta z_i(t)) \\
&\le \frac{1}{2}\sum_{i=1}^{N} \Delta z_i^T(t)(A_i + A_i^T + B_i^T B_i + h^2C^TC - 2aI_n)\Delta z_i(t) \\
&\quad + b\sum_{i=1}^{N} \Delta z_i^T(qt)\Delta z_i(qt) + \sum_{i=1}^{N} l_i y_i^2 + \sum_{k=1}^{n} \tilde{\eta}_k^T(t)\Pi\tilde{\eta}_k(t) \\
&\le -\alpha V(t) + \beta V(qt) + \sum_{i=1}^{N} l_i y_i^2,
\end{aligned} \tag{5.16}$$

where $\tilde{Y}_k(\tilde{z}(t)) = [Y_1(z^k(t)), Y_2(z^k(t)), \ldots, Y_N(z^k(t))]^T$ for $k = 1, 2, \ldots, n$, $\eta_k(t) = [\Delta\tilde{z}_k(t)^T, \Delta\tilde{z}_k(qt)^T, \tilde{Y}_k(\tilde{z}(t))^T]^T, \Delta\tilde{z}_k(t) = [\Delta z_1^k(t), \Delta z_2^k(t), \ldots, \Delta z_N^k(t)]^T$ and $L = \mathrm{diag}\{l_1, l_2, \ldots, l_N\}$.

Considering the following impulsive comparison system with a unique solution $\gamma(t)$ for any $\varepsilon > 0$ and combining the inequities (5.15) with (5.16), one can derive that

$$\begin{cases}
\dot{\gamma}(t) = -\alpha\gamma(t) + \beta\gamma(qt) + \sum_{i=1}^{N} l_i y_i^2 + \varepsilon, \ t \neq t_k, k \in \mathbb{N}, \\
\gamma(t_k^+) = \omega\gamma(t_k^-), \ t = t_k, k \in \mathbb{N}, \\
\gamma(t) = \sum_{i=1}^{N} \|\psi_i(0)\|^2 = \|\psi(0)\|.
\end{cases} \tag{5.17}$$

In view of the comparison principle in Lemma 5.1, we could obtain that $V(t) \le \gamma(t)$ for any $t > 0$. Due to the extended formula for the variation of parameters [36], one could get the following integral equation for $\gamma(t)$ with proportional time-varying delay $\gamma(qt)$

$$\gamma(t) = \Phi(t,0)\gamma(0) + \int_0^t \Phi(t,s)(\beta\gamma(qs) + \sum_{i=1}^{N} l_i y_i^2 + \varepsilon)ds, \tag{5.18}$$

where $\Phi(t,s)(t \geq s \geq 0)$ is the Cauchy matrix of the following linear impulsive system

$$\begin{cases} \dot{\gamma}(t) = -\alpha\gamma(t), & t \neq t_k, k \in \mathbb{N}, \\ \gamma(t_k^+) = \omega\gamma(t_k^-), & t = t_k, k \in \mathbb{N}. \end{cases}$$

Case I. For $0 < \omega \leq 1$, the Cauchy matrix $\Phi(t,s)$ can be calculated with the concept of average impulsive interval $N_\zeta(t,s)$ in right side of Definition 5.2 as follows

$$\begin{aligned} \Phi(t,s) &= e^{-\alpha(t-s)} \prod_{s \leq t_k \leq t} \omega \\ &\leq e^{-\alpha(t-s)} \omega^{N_\zeta(t,s)} \\ &\leq e^{-\alpha(t-s)} \omega^{\frac{t-s}{N_a} - N_0} \\ &\leq \omega^{-N_0} e^{(\frac{\ln\omega}{N_a} - \alpha)(t-s)}. \end{aligned} \tag{5.19}$$

Substituting (5.19) into the integral equation (5.18) gives that

$$\begin{aligned} \gamma(t) &= \Phi(t,0)\gamma(0) + \int_0^t \Phi(t,s)\left[\beta\gamma(qs) + (\sum_{i=1}^{N} l_i y_i^2 + \varepsilon)\right]ds \\ &\leq \omega^{-N_0} \sum_{i=1}^{N} \|\psi_i(0)\|^2 e^{(\frac{\ln\omega}{N_a} - \alpha)t} + \int_0^t \omega^{-N_0} e^{(\frac{\ln\omega}{N_a} - \alpha)(t-s)} \\ &\quad \times (\beta\gamma(qs) + (\sum_{i=1}^{N} l_i y_i^2 + \varepsilon))ds \\ &= \omega_0 e^{\xi t} + \int_0^t e^{\xi(t-s)} \omega^{-N_0} (\beta\gamma(qs) + (\sum_{i=1}^{N} l_i y_i^2 + \varepsilon))ds, \end{aligned} \tag{5.20}$$

where $\omega_0 = \omega^{-N_0} \sum_{i=1}^{N} \|\psi_i(0)\|^2$ and $\xi = \frac{\ln\omega}{N_a} - \alpha$.

In the next, based on analysis in inequality (5.20), the exponential estimation of $\gamma(t)$ will be derived by introducing the mathematical contradiction method. Meanwhile, it is necessary that $\xi = \frac{\ln\omega}{N_a} - \alpha < 0$ for this purpose.

From the above information, if there exists λ meeting (5.10), there holds

$$\gamma(t) < \omega_0 e^{-\lambda t} - \frac{\omega^{-N_0}(\sum_{i=1}^{N} l_i y_i^2 + \varepsilon)}{\xi + \beta\omega^{-N_0}}, \quad t > 0. \tag{5.21}$$

Specially, there exists $\gamma(0) < \omega_0 - \frac{\omega^{-N_0}(\sum_{i=1}^{N} l_i y_i^2 + \varepsilon)}{\xi + \beta\omega^{-N_0}}$ for $t = 0$.

Then, with the analytic method of contradiction, the effectiveness of (5.21) will be illustrated. When the assumption is false for all $t > 0$, which implies there are at least one time instant $t^* > 0$ meeting

$$\gamma(t^*) \geq \omega_0 e^{-\lambda t^*} - \frac{\omega^{-N_0}(\sum_{i=1}^N l_i y_i^2 + \varepsilon)}{\xi + \beta\omega^{-N_0}}, t^* > t > 0, \tag{5.22}$$

but for $0 < t < t^*$, there still holds (5.21). Therefore, due to (5.20) and (5.21), (5.22), the following inequalities could be achieved

$$\begin{aligned}
\gamma(t^*) &\leq \omega_0 e^{\xi t^*} + \int_0^{t^*} e^{\xi(t^*-s)}[\beta\omega^{-N_0}\gamma(qs) + \omega^{-N_0}(\sum_{i=1}^N l_i y_i^2 + \varepsilon)]ds \\
&< \omega_0 e^{\xi t^*} + \int_0^{t^*} e^{\xi(t^*-s)}\beta\omega^{-N_0}(\omega_0 e^{-\lambda qs} - \frac{\omega^{-N_0}(\sum_{i=1}^N l_i y_i^2 + \varepsilon)}{\xi + \beta\omega^{-N_0}})ds \\
&\quad + \int_0^{t^*} e^{\xi(t^*-s)}\omega^{-N_0}(\sum_{i=1}^N l_i y_i^2 + \varepsilon)ds \\
&= \omega_0 e^{\xi t^*} + \beta\omega^{-N_0}\omega_0 e^{\xi t^*}\int_0^{t^*} e^{-(\xi+\lambda q)s}ds + (\beta\omega^{-N_0}\theta \\
&\quad + \omega^{-N_0}(\sum_{i=1}^N l_i y_i^2 + \varepsilon))\int_0^{t^*} e^{\xi(t^*-s)}ds \\
&= \omega_0 e^{\xi t^*} - \frac{\beta\omega^{-N_0}\omega_0}{\xi + \lambda q} e^{\xi t^*}(e^{-(\xi+\lambda q)t^*} - 1) \\
&\quad + \frac{\omega^{-N_0}(\beta\theta + (\sum_{i=1}^N l_i y_i^2 + \varepsilon))}{-\xi} e^{\xi t^*}(e^{-\xi t^*} - 1) \\
&= \omega_0 e^{\xi t^*} - \frac{\beta\omega_0\omega^{-N_0}}{\xi + \lambda q}(e^{-\lambda q t^*} - e^{\xi t^*}) \\
&\quad + \frac{\omega^{-N_0}(\beta\theta + (\sum_{i=1}^N l_i y_i^2 + \varepsilon))}{-\xi}(1 - e^{\xi t^*}),
\end{aligned} \tag{5.23}$$

where $\theta = -\frac{\omega^{-N_0}(\sum_{i=1}^N l_i y_i^2 + \varepsilon)}{\xi + \beta\omega^{-N_0}}$. Therefore, there holds $\xi\theta + \beta\omega^{-N_0}\theta = -\omega^{-N_0}(\sum_{i=1}^N l_i y_i^2 + \varepsilon)$, that is, $\xi\theta = -\omega^{-N_0}(\sum_{i=1}^N l_i y_i^2 + \varepsilon + \beta\theta)$. Therefore, (5.23) can be further calculated as

$$\gamma(t^*) \leq \omega_0 e^{\xi t^*} - \frac{\beta\omega_0\omega^{-N_0}}{\xi + \lambda q}(e^{-\lambda q t^*} - e^{\xi t^*}) + \theta(1 - e^{\xi t^*}). \tag{5.24}$$

The following procedure is presented to show the contradiction from the above conclusion (5.24). Define the parameter function as

$$\tilde{s}(t) = \xi(\xi + \lambda q + \beta\omega^{-N_0})e^{\xi t} + \lambda(\xi + \lambda q + q\beta\omega^{-N_0})e^{-\lambda t}. \tag{5.25}$$

Calculating the derivative of $\tilde{s}(t)$ gives

$$\dot{\tilde{s}}(t) = \xi^2(\xi + \lambda q + \beta\omega^{-N_0})e^{\xi t} - \lambda^2(\xi + \lambda q + q\beta\omega^{-N_0})e^{-\lambda t}. \tag{5.26}$$

Furthermore, $\dot{\tilde{s}}(t) = 0$ if and only if the time instant t in (5.26) is

$$t = T_s = \frac{1}{\xi + \lambda} \ln \frac{\lambda^2((\xi + \lambda q) + q\beta\omega^{-N_0})}{\xi^2((\xi + \lambda q) + \beta\omega^{-N_0})},$$

where exists $0 < \frac{\lambda^2((\xi+\lambda q)+q\beta\omega^{-N_0})}{\xi^2((\xi+\lambda q)+\beta\omega^{-N_0})} < 1$ from the condition (5.10). There holds $\dot{\tilde{s}}(t) > 0$ for $0 < t < T_s$, which implies $\tilde{s}(t)$ is increasing for $0 < t < T_s$. In the meantime, there holds $\dot{\tilde{s}}(t) < 0$ for $t > T_s$, which denotes that $\tilde{s}(t)$ is decreasing when $t > T_s$. Additionally, there holds $\tilde{s}(0) = \lambda[(\xi + \lambda q) + q\beta\omega^{-N_0}] + \xi[(\xi + \lambda q) + \beta\omega^{-N_0}] > 0$ through analysis in condition (5.10). Especially, $\lim_{t\to+\infty} \tilde{s}(t) = 0$ with $\Delta < 0$ from (5.25). Therefore, $\tilde{s}(t) \geq 0$ for $t \geq 0$ can be deduced.

Since $\xi + \lambda q < 0$ and $\tilde{s}(t) \geq 0$, then one obtains

$$\xi(1 + \frac{\beta\omega^{-N_0}}{\xi + \lambda q})e^{\xi t} + \lambda(1 + \frac{q\beta\omega^{-N_0}}{\xi + \lambda q})e^{-\lambda t} \leq 0. \tag{5.27}$$

According to $0 < q < 1$ and $\lambda > 0$, we could achieve $e^{-\lambda q t} > e^{-\lambda t}$. Then, based on the above analysis in (5.27), the inequality can be concluded as follows

$$\Upsilon(t) \triangleq \xi(1 + \frac{\beta\omega^{-N_0}}{\xi + \lambda q})e^{\xi t} + \lambda\frac{q\beta\omega^{-N_0}}{\xi + \lambda q}e^{-\lambda q t} + \lambda e^{-\lambda t} \leq 0. \tag{5.28}$$

Define another parameter equation $s(t)$ as

$$s(t) = (1 + \frac{\beta\omega^{-N_0}}{\xi + \lambda q})\, e^{\xi t} - \frac{\beta\omega^{-N_0}}{\xi + \lambda q}\, e^{-\lambda q t} - e^{-\lambda t}. \tag{5.29}$$

One can derive that $s(0) = 0$ and $\dot{s}(t) = \Upsilon(t)$, which means that $\dot{s}(t) \leq 0$ and $s(t)$ is a monotone decreasing function with initial value $s(0) = 0$. Therefore, there holds $s(t) \leq s(0) = 0$, that is,

$$(1 + \frac{\beta\omega^{-N_0}}{\xi + \lambda q})\, e^{\xi t} - \frac{\beta\omega^{-N_0}}{\xi + \lambda q}\, e^{-\lambda q t} \leq e^{-\lambda t}, \qquad t \geq 0. \tag{5.30}$$

Considering the inequalities (5.24) and (5.30), it gives

$$\begin{aligned}
\gamma(t^*) &\le \omega_0(1+\frac{\beta\omega^{-N_0}}{\xi+\lambda q})\, e^{\xi t^*} - \omega_0 \frac{\beta\omega^{-N_0}}{\xi+\lambda q}\, e^{-\lambda q t^*} + \theta(1-e^{\xi t^*}) \\
&\le \omega_0\, e^{-\lambda t^*} + \theta(1-e^{\xi t^*}) \\
&< \omega_0\, e^{-\lambda t^*} + \theta \\
&= \omega_0 e^{-\lambda t^*} - \frac{\omega^{-N_0}(\sum_{i=1}^N l_i y_i^2 + \varepsilon)}{\xi + \beta\omega^{-N_0}},
\end{aligned} \tag{5.31}$$

which results in a contradiction to (5.22). It is proved that Assumption 5.1 is effective for $t \ge 0$. In view of the comparison principle, it obtains

$$0 < V(t) \le \gamma(t) < \omega_0 e^{-\lambda t} - \frac{\omega^{-N_0}(\sum_{i=1}^N l_i y_i^2 + \varepsilon)}{\xi + \beta\omega^{-N_0}}.$$

For $\varepsilon \to 0$, it gives

$$\|\Delta z(t)\| \le \sqrt{2\omega^{-N_0}\sum_{i=1}^{N}\|\psi_i(0)\|\, e^{-\frac{\lambda}{2}t}} + \sqrt{\frac{2\omega^{-N_0}\sum_{i=1}^N l_i y_i^2}{-(\xi+\beta\omega^{-N_0})}}.$$

According to the above estimation of the error vector $\Delta z(t)$, one could obtain that there exists a compact set

$$\bar{C} = \left\{ \Delta z(t) \in \mathbb{R}^{nN} \,|\, \|\Delta z(t)\| \le \bar{z} = \sqrt{\frac{2\omega^{-N_0}\sum_{i=1}^N l_i y_i^2}{-(\xi+\beta\omega^{-N_0})}} \right\},$$

where $\bar{z}$ is the synchronization error bound and t goes to infinity. In addition, from above analysis, we can draw the conclusion that the error Lur'e networks (5.6) exponentially converge into the compact set $\bar{C}$ with the convergence rate $\frac{\lambda}{2}$. Until now, for $0 < \omega \le 1$, the synchronization between the Lur'e networks (5.1) and the target Lur'e system (5.2) is finally achieved under the distributed pinning controller (5.5) within the synchronization error bound $\bar{z}$.

Case II. For the condition $\omega > 1$, the Cauchy matrix $\Phi(t,s)$ can also be estimated by the concept of average impulsive interval

$$\begin{aligned}
\Phi(t,s) &= e^{-\alpha(t-s)} \prod_{s \le t_k \le t} \omega \le e^{-\alpha(t-s)} \omega^{N_\zeta(t,s)} \\
&\le e^{-\alpha(t-s)} \omega^{\frac{t-s}{N_a}+N_0} \le \omega^{N_0} e^{(\frac{\ln\omega}{N_a}-\alpha)(t-s)}.
\end{aligned} \tag{5.32}$$

With the same method, we could correspondingly achieve

$$\gamma(t) \leq \bar{\omega}_0 e^{\xi t} + \int_0^t e^{\xi(t-s)} \left[\beta \omega^{N_0} \gamma(qs) + \omega^{N_0} (\sum_{i=1}^{N} l_i y_i^2 + \varepsilon) \right] ds, \tag{5.33}$$

where $\bar{\omega}_0 = \omega^{N_0} \sum_{i=1}^{N} \|\psi_i(0)\|^2$.

According to the analysis in **Case I**, if there exists a positive constant $\bar{\lambda}$ which satisfies (5.12), the following inequality could be proved

$$\gamma(t) < \bar{\omega}_0 e^{-\bar{\lambda} t} - \frac{\omega^{N_0}(\sum_{i=1}^{N} l_i y_i^2 + \varepsilon)}{\xi + \beta \omega^{N_0}}, \quad t > 0.$$

With the same steps, we deduce

$$0 < V(t) \leq \gamma(t) < \bar{\omega}_0 e^{-\bar{\lambda} t} - \frac{\omega^{N_0}(\sum_{i=1}^{N} l_i y_i^2 + \varepsilon)}{\xi + \beta \omega^{N_0}}.$$

Similarly, for $\varepsilon \to 0$, the following inequality holds

$$\|\Delta z(t)\| \leq \sqrt{2\omega^{N_0} \sum_{i=1}^{N} \|\psi_i(0)\|\, e^{-\frac{\bar{\lambda}}{2} t}} + \sqrt{\frac{2\omega^{N_0} \sum_{i=1}^{N} l_i y_i^2}{-(\xi + \beta \omega^{N_0})}}.$$

Therefore, different from **Case I**, the compact set $\bar{C}$ could be rewritten as follows

$$\bar{C} = \left\{ \Delta z(t) \in \mathbb{R}^{nN} | \|\Delta z(t)\| \leq \tilde{z} = \sqrt{\frac{2\omega^{N_0} \sum_{i=1}^{N} l_i y_i^2}{-(\xi + \beta \omega^{N_0})}} \right\},$$

Through similar proof procedure, one can derive that the error Lur'e networks (5.6) exponentially converge into the compact set $\bar{C}$ with the convergence rate $\frac{\bar{\lambda}}{2}$. That is, the quasi-synchronization between the Lur'e networks (5.1) and (5.2) is achieved for $\omega > 1$ under the distributed pinning controller (5.5) within the synchronization error bound $\tilde{z}$. Until now, the proof of the theorem is totally completed. □

Remark 5.2 In consideration of the quasi-synchronization of nonlinear Lur'e networks with proportional delay. From the Lur'e network (5.1), the dynamics at time instant t is subject to its current states $z_i(t)$ and $z_i(qt)$ with proportional time rate, where time ratio $q \in (0, 1)$. Comparing with normal time invariant delay τ or time-varying delay $\tau(t)$, proportional delay discussed in this chapter is a kind of unbounded time-varying delays due to a prescribed q and the term $qt \to +\infty$ as $t \to \infty$. Therefore, derivation procedure of the exponential estimation of (5.20) becomes more difficult and challengeable, which is different with the previous works [10, 16, 17]. The main reason behind it is that the time evolution of the nonidentical Lur'e network, rather than linearly transformed by normal time unit, is proportionally shifted in the term qt with the proportional factor. In order to settle this problem, a new comparison system has been constructed which contains differential inequalities and system heterogeneities. And the formula for the variation of parameters has been also applied in the proof. Moreover, some parameter equations like (5.25), (5.28) and (5.29) are artificially given to achieve the contradiction to the inequality (5.22).

Remark 5.3 Generally, as a kind of superior discontinuous control protocols, impulsive control supplies an instant motion power to the controlled systems, which largely saves energy rather than previous continuous control methods. Focusing on the impulsive problem, impulse could be either viewed as the disturbances or the control input which implies that the impulsive effect μ plays an important role in the synchronization. It is clear that many scholars have discussed the synchronization of complex networks with positive impulsive effects in previous works [17, 32–35]. However, the synchronization of complex networks with negative impulsive effects should also be discussed. Therefore, in this chapter, the designed distributed impulsive pinning controller (5.5) contains the impulsive control part. In view of the different functions of impulsive effects in this controller, different conditions on the selection strategy of parameter ω are studied. If impulsive effect μ plays a positive role in the synchronization, that is $0 < \omega \leq 1$, the feedback control gains d_i could spend small amounts of control costs for the first situation. And for the second situation $\omega > 1$, where the impulsive effect μ hampers the synchronization of the complex networks, or even causes the unstable or damage to the networks. In this situation, the negative feedback control term $-d_i \Delta z_i(t)$ functions effectively to counteract the side-effects brought by the disadvantageous impulse. The situations will be illustrated in detail by numerical simulations in the next section.

Remark 5.4 From the above discussion, different criteria for quasi-synchronization of the nonidentical Lur'e networks have been realized with respect to different situations of the parameter ω. Actually, a direct relationship between the parameter ω and the impulsive effect μ can be easily found in the LMI (5.8). In the proof procedure, the inequality $(\mu(kW - D) \otimes I_n + I_{Nn})^T(\mu(kW - D) \otimes I_n + I_{Nn}) < \omega I_{Nn}$ finally results in the satisfaction of (5.15). According to the different kind of impulsive effects, the parameter ω could be calculated to judge which situation it is. Some numerical examples with different values of impulsive effect will be presented to explain this problem.

Remark 5.5 Through adjusting the freedom indicator N_0 and joint utilization of the positive number N_a, the number of impulses during the interval (t, T) could be estimated based on the concept of average impulsive interval. Especially, N_a has been decided as $N_a = \min_k\{t_k - t_{k-1}\}$ or $N_a = \max_k\{t_k - t_{k-1}\}$ in some previous works [33, 37]. One can conclude that this kind of definition on N_a may lead to the final conservative results due to the positive number N_a satisfying the condition $\min_k\{t_k - t_{k-1}\} \leq N_a \leq \max_k\{t_k - t_{k-1}\}$. For instance, with the definition of previous works and the condition $\tilde{N}_a = \max_k\{t_k - t_{k-1}\}$, the inequality of condition (5.10) should be rewritten as $\frac{\ln \omega}{\tilde{N}_a} - a + \lambda q + \beta \omega^{-N_0} < 0$. Then a smaller $\tilde{N}_a$ could satisfy the condition of this chapter for the same given parameters. In other words, in order to realize the synchronization between the nonlinear Lur'e systems, it will be best to improve control cost. In this regard, the concept of the average impulsive interval is introduced to effectively decrease the conservatism and largely save the control costs in this chapter.

Remark 5.6 The delay independent synchronization criteria could be obtained and there is no specific value required for time ratio q in the unbounded delay term $z_j(qt)$ ahead of time from (5.8) and (5.9) in Theorem 5.1. However, from the definition of convergence rate of the exponential synchronization $\frac{\lambda}{2}$, one can find that the proportional factor influences the convergence rate of the final synchronization. Additionally, the inequalities in (5.10) show that the convergence rate of synchronization improved along with the bigger time ratio q.

Remark 5.7 Until now serval classical methods have been utilized to estimate the synchronization error bound. Firstly, based on the construction of generalized Halanay inequalities of Volterra functional differential equations, the synchronization error has been obtained in [16]. Secondly, in [13], the synchronization error bound has been estimated by computing the upper bound of the corresponding Lyapunov function directly. Thirdly, after analyzing the minimum eigenvalue of matrix, the synchronization error has been evaluated via constructing a new Lyapunov functional equation $\tilde{V}(t, e(t)) = e^{\sigma t}V(t, e(t))$ which is the product of the exponent term $e^{\sigma t}$ and normal Lyapunov functional $V(t, e(t))$ in [12]. And in this chapter, by combining the comparison system (5.17), the formula of the variation of parameters and the concept of the average impulsive interval, the upper bound of $\gamma(t)$ has been estimated and the synchronization errors has been achieved finally.

5.4 Numerical Simulations

In this section, some numerical simulations will be presented to prove the validity of theoretical result in above discussion. Consider the following Chua's circuits with prescribed parameters

$$\begin{cases} \dot{z}^1(t) = a_k(z^2(t) - h(z^1(t))), \\ \dot{z}^2(t) = z^1(t) - z^2(t) + z^3(t), \\ \dot{z}^3(t) = -b_k z^2(t) - c_k z^3(t), \end{cases} \tag{5.34}$$

where nonlinear function $h(z^1(t))$ could be denoted as $h(z^1(t)) = q_k z^1(t) + \frac{1}{2}(p_k - q_k)(|z^1(t) + 1| - |z^1(t) - 1|)$ for $k = 1, 2, 3$ with $a_1 = 9.78, b_1 = 14.97,\ c_1 = 0,$ $p_1 = 1.31,\ q_1 = 0.75;\ a_2 = 10, b_2 = 14.87,\ c_2 = 0, p_2 = 1.27, q_2 = 0.68;\ a_3 = 10, b_3 = 15, c_3 = 0.0385, p_3 = 1.27, q_3 = 0.68$.

Combining the above Chua's circuits and Lur'e forms, the Lur'e network could be rewritten as

$$\dot{z}(t) = A_k z(t) + B_k \tilde{f}(Cz(t)), \tag{5.35}$$

where

$$\tilde{f}(z^1(t)) = \frac{1}{2}(p_k - q_k)(|z^1(t) + 1| - |z^1(t) - 1|),$$

$$A_k = \begin{bmatrix} -a_k q_k & a_k & 0 \\ 1 & -1 & 1 \\ 0 & -b_k & -c_k \end{bmatrix}, \; B_k = \begin{bmatrix} -a_k(p_k - q_k) \\ 0 \\ 0 \end{bmatrix}, \; C = \begin{bmatrix} 1 \, 0 \, 0 \end{bmatrix}$$

for $k = 1, 2, 3$. Then define system parameters of the target Lur'e system (5.2) as $a_1 = 9.78, b_1 = 14.97, c_1 = 0, p_1 = 1.31, q_1 = 0.75$. The coupled Lur'e network (5.1) consisting of three different Lur'e systems are described in (5.35). And therefore, our goal is to synchronize the three coupled Lur'e systems to the target Lur'e system within a prescribed synchronization error bound $\bar{z}$.

Considering the topology structure of coupled Lur'e dynamical network, let the coupling matrix $G = [1, -1, 0; -1, 2, -1; 0, -1, 0]$, the coupling strength $c = 0.2$ and the factor $q = 0.8$ in proportional time-varying delay. Especially, one could notice that the delay of coupling term $c\sum_{j=1}^{N} g_{ij}\Gamma z_j(qt)$ is a class of unbounded time-varying delays. In view of the distributed impulsive pinning controller (5.5), set the average impulsive interval $N_a = 0.02$ and the free adjust constant $N_0 = 1$. The synchronization error of the three states between coupled Lur'e networks and the target Lur'e systems is defined as $E(t) = \sqrt{\frac{1}{nN}\sum_{i=1}^{N}\sum_{j=1}^{n}(\Delta z_i^j(t))^2}$.

In order to synchronize the coupled Lur'e network (5.1) and target Lur'e system (5.23), the distributed impulsive pinning controller (5.5) will be fully utilized. In the next, the influences brought by the control gain and impulsive effect will be presented on the final quasi-synchronization of the Lur'e networks.

Example 5.1 In view of different functions of the impulsive effect, choose $\mu = 0.8$, $k = 0.5$ and feedback control gain $d_i = 4.55$, $i = 1, 2, 3$ in the first example. For the above information, $\omega = 2.2 > 1$ is effective for the condition (5.8) in Theorem 5.1. Through selecting some suitable constants a and b, one can verify that condition (ii) in Theorem 5.1 is satisfied. Choose $\alpha = 1000$, $\beta = 10$, one could notice that $\frac{\ln\omega}{N_a} - \alpha + \beta\omega^{-N_0} = -938.5771 < 0$, that is condition (iii-b) in Theorem 5.1 is satisfied. Then, with the above information, the calculation on the synchronization error bound $\bar{z}$ yields that

$$\bar{z} = \sqrt{\frac{2\omega^{N_0}\sum_{i=1}^{N} l_i y_i^2}{-(\frac{\ln\omega}{N_a} - \alpha + \beta\omega^{N_0})}} = 0.0876.$$

In this respect, the bound synchronization between the coupled Lur'e networks (5.1) and target Lur'e system (5.2) is successfully realized with the prescribed compact set $\bar{C} = \{\Delta z(t) \in R^{nN} \mid \|\Delta z(t)\| \le \bar{z} = 0.0876\}$.

Figure 5.1a–c plot the states of the three Lur'e systems under the distributed impulsive controller (5.5). Obviously, the above pictures explain that three states of the Lur'e systems are synchronized within the bound. And the evolution curve of

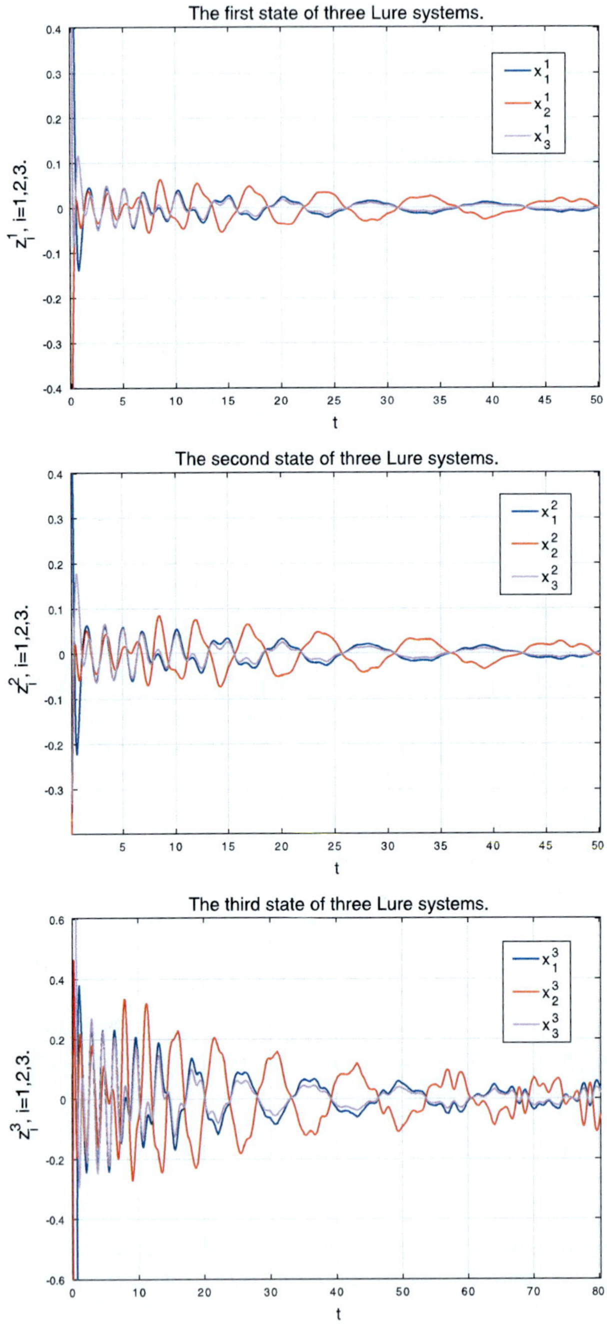

Fig. 5.1 The evolution curves of three states of Lur'e systems

the synchronization error $E(t)$ is plotted in Fig. 5.2a, where the curve shows that the practical error is less than the theoretical synchronization error bound $\bar{z}$. Figure 5.2b shows three coupled Lur'e systems.

Example 5.2 From the investigation described above, the different impulsive effects and different values of ω were discussed, respectively. Therefore, set $\mu = -0.8$, $k = 6$ to show this point in the second example. For the given information, $\omega = 0.7276 < 1$ can be calculated. Similar to Example 5.1, the purpose of condition (ii) with the selected scalars a and b in Theorem 5.1 is finished. Choose $\alpha = 50$, $\beta = 10$ based on the selected parameters, there holds $\frac{\ln \omega}{N_a} - \alpha + \beta\omega^{-N_0} = -52.1564 < 0$ satisfying the condition (iii-b). Then, the following bound $\bar{z}$ could be calculated as

$$\bar{z} = \sqrt{\frac{2\omega^{-N_0}\sum_{i=1}^{N} l_i y_i^2}{-(\frac{\ln \omega}{N_a} - \alpha + \beta\omega^{-N_0})}} = 0.0863.$$

In view of the synchronization error bound $\bar{z}$, one could draw the conclusion that the quasi-synchronization between the coupled Lur'e networks (5.1) and target Lur'e system (5.2) is successfully achieved with the prescribed compact set $\bar{C} = \{\Delta z(t) \in R^{nN} \mid \|\Delta z(t)\| \leq \bar{z} = 0.0863\}$.

Moreover, Fig. 5.3 expresses the error curves with the definition of synchronization error, where the experiential error is less than the theoretical synchronization error bound $\bar{z}$.

Example 5.3 In this example, in view of the distributed impulsive control term and the distributed impulsive controller (5.5) without negative feedback control input, define $\mu = 0.8$, but $d_1 = d_2 = d_3 = 0$. It can be verified that the synchronization error $E(t)$ will continuously increase. It further denotes that the quasi-synchronization cannot be achieved under the distributed pinning impulsive controller (5.5) with $d_1 = d_2 = d_3 = 0$, which explains the conditions in Remark 5.3.

5.5 Conclusion

This chapter has studied the quasi-synchronization of nonlinear coupled Lur'e networks with the proportional delay and mismatched parameters. As a type of unbounded time-varying delays, the proportional delay leads to tremendously increased the difficulty on synchronization, which is different from general time delays. Therefore, a distributed impulsive pinning controller has been designed to solve this kind of problem. For different effects caused by different impulsive effects in network synchronization, sufficient criteria have been proposed based on the delayed impulsive comparison principle, the extended formula for the variation of parameters and the concept of an average impulsive interval. In addition, synchronization errors and exponential convergence rates have been calculated simultaneously

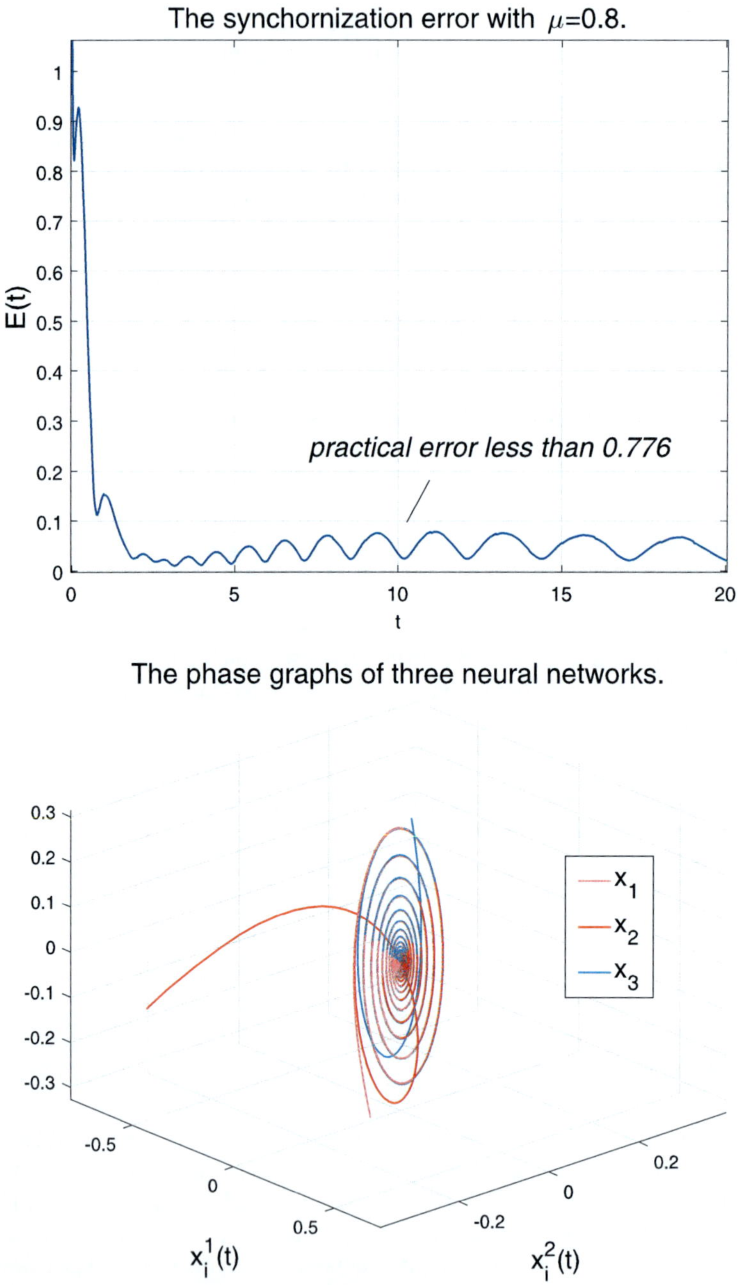

Fig. 5.2 The state error curves of the Lur'e networks with $\mu = 0.8$, $d_i = 4.55$, $i = 1, 2, 3$, and the phase graphs of three Lur'e systems with different system parameters

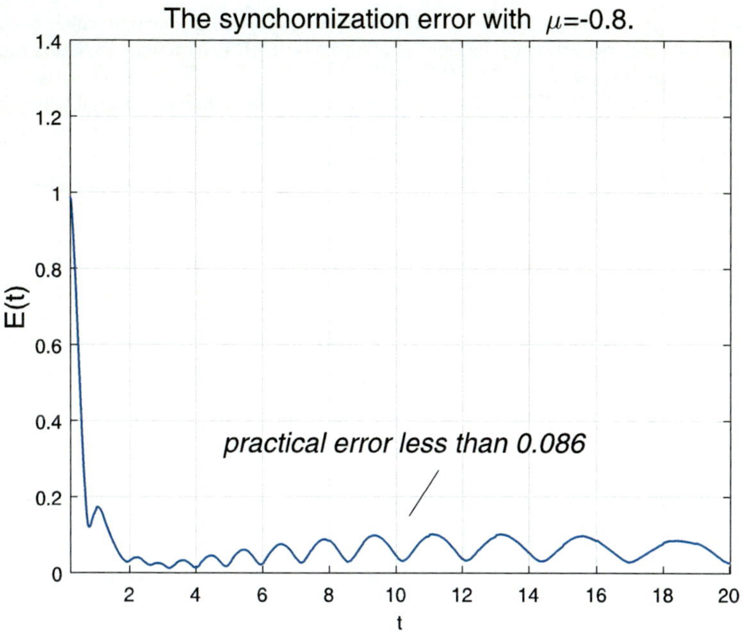

Fig. 5.3 The state error curves of the Lur'e networks with $\mu = -0.8, d_i = 1.6, i = 1, 2, 3$

for different impulsive effects with different functions. Finally, three numerical simulations have been presented to prove the validity of the above theoretical results. Different synchronization problems of complex networks involving general time-varying delay, distributed delay or proportional delay could be further studied in our following work.

References

1. Strogatz, S. H.: Exploring complex networks. Nature. **410**, 268–276 (2001)
2. Liu, Y. Y., Slotine, J. J., Barabási, A. L.: Controllability of complex networks. Nature. **473**, 167–173 (2011)
3. Benson, A. R., Gieich, D. F., Leskovec, J.: Higher-order organization of complex networks. Science. **353**, 163–166 (2016)
4. Li, C. G., Chen, G. R.: Synchronization in general complex dynamical networks with coupling delays. Physica A: Statistical Mechanics and its Applications. **343**, 263–278 (2004)
5. Liu, X. W.: Synchronization analysis for nonlinearly-coupled complex networks with an asymmetrical coupling matrix. Physica A: Statistical Mechanics and its Applications. **387**, 4429–4439 (2008)
6. Wu, Z. G., Shi, P., Su, H. Y., Chu, J.: Stochastic synchronization of Markovian jump neural networks with time-varying delay using sampled-data. IEEE Transactions on Cybernetics. **43**, 1796–1806 (2013)

7. Wu, Z. G., Shi, P., Su, H. Y., Chu, J.: Sampled-data exponential synchronization of complex dynamical networks with time-varying coupling delay. IEEE Transactions on Neural Networks and Learning Systems. **24**, 1177–1187 (2013)
8. Wu, Z. G., Shi, P., Su, H. Y., Chu, J.: Exponential synchronization of neural networks with discrete and distributed delays under time-varying sampling. IEEE Transactions on Neural Networks and Learning Systems. **23**, 1368–1376 (2012)
9. Zhang, S., Yu, Y. G., Wen, G. G., Rahmani, A.: Stochastic quasi-synchronization for uncertain chaotic delayed neural networks. International Journal of Modern Physics C. **25**, 1450029 (2014)
10. Tang, Z., Park, J. H., Feng, J. W.: Impulsive effects on quasi-synchronization of neural networks with parameter mismatches and time-varying delay. IEEE Transactions on Neural Networks and Learning Systems. **29**, 908–919 (2018)
11. Liu, X. Y., Chen, T. P., Cao, J. D., Lu, W. L.: Dissipativity and quasi-synchronization for neural networks with discontinuous activations and parameter mismatches, Neural Networks. **24**, 1013–1021 (2011)
12. He, W. L., Qian, F., Han, Q.-L., Cao, J. D.: Synchronization error estimation and controller design for delayed Lur'e systems with parameter mismatches, IEEE Transactions on Neural Networks and Learning Systems. **23**, 1551–1563 (2012)
13. Liu, C., Li, C. D., Li, C. J.: Quasi-synchronization of delayed chaotic systems with parameters mismatch and stochastic perturbation. Communications in Nonlinear Science and Numerical Simulation. **16**, 4108–4119 (2011)
14. Pan, L. J., Cao, J. D.: Stochastic quasi-synchronization for delayed dynamical networks via intermittent control. Communications in Nonlinear Science and Numerical Simulation. **17**, 1332–1343 (2012)
15. Nishio, Y., Ushida, A.: Quasi-synchronization phenomena in chaotic circuits coupled by one resistor. IEEE Transactions on Circuits Systems I: Fundamental Theory and Applications. **43**, 491–496 (1996)
16. He, W. L., Qian, F., Han, Q.-L., Cao, J. D.: Lag quasi-synchronization of coupled delayed systems with parameter mismatch. IEEE Transactions on Circuits Systems I: Regular Papers. **58**, 1345–1357 (2011)
17. He, W. L., Qian, F., Lam, J., Chen, G. R., Han, Q.-L., Kurths, J.: Quasi-synchronization of heterogeneous dynamic networks via distributed impulsive control: Error estimation, optimization and design. Automatica. **62**, 249–262 (2015)
18. Wu, Z.-G., Shi, P., Su, H. Y., Chu, J.: Stochastic synchronization of Markovian jump neural networks with time-varying delay using sampled data. IEEE Transactions on Cybernetics. **43**, 1796–1806 (2013)
19. Zhong, J., Lu, J. Q., Liu, Y., Cao, J. D.: Synchronization in an array of output-coupled Boolean networks with time delay. IEEE Transactions on Neural Networks and Learning Systems. **25**, 2288–2294 (2014)
20. Lu, J. Q., Zhong, J., Ho, Daniel W. C., Tang, Y., Cao, J. D.: On controllability of delayed Boolean control networks. SIAM Journal on Control and Optimization. **54**, 475–494 (2016)
21. Wang, L. M., Shen, Y., Zhang, G. D.: Synchronization of a class of switched neural networks with time-varying delays via nonlinear feedback control. IEEE Transactions on Cybernetics. **46**, 2300–2310 (2016)
22. Yang, X. S., Ho, Daniel W. C.: Synchronization of delayed memrisitive neural networks: robust analysis approach. IEEE Transactions on Cybernetics. **46**, 3377–3387 (2016)
23. Zhou, L. Q., Zhang, Y. Y.: Global exponential stability of cellular neural networks with multi-proportional delays. International Journal of Biomathematics. **8**, 1550071 (2015)
24. Zhou, W. N., Wang, B. T., Zhong, Q.-C., Fang, J. A.: Proportional-delay adaptive control for global synchronization of complex networks with time-delay and switching outer-coupling matrices. International Journal of Robust and Nonlinear Control. **23**, 548–561 (2013)
25. Zheng, C., Li, N., Cao, J. D.: Matrix measure based stability criteria for high-order neural networks with proportional delay. Neurocomputing. **149**, 1149–1154 (2015)

26. Jia, R. W.: Finite-time stability of a class of fuzzy cellular neural networks with multi-proportional delays. Fuzzy Sets and Systems. **319**, 70–80 (2017)
27. Zhou, L. Q.: Delay-dependent exponential synchronization of recurrent neural networks with multiple proportional delays. Neural Processing Letters. **42**, 619–632 (2015)
28. Zhou, L. Q., Chen, X. B., Yang, Y. X.: Asymptotic stability of cellular neural networks with multiple proportional delays. Computers and Mathematics with Applications. **229**, 457–466 (2014)
29. Chen, Y., Yu, W. W., Li, F. F., Feng, S. S.: Synchronization of complex networks with impulsive control and disconnected topology. IEEE Transactions Circuits Systems II: Express Briefs. **60**, 292–296 (2013)
30. Xu, W. Y., Ho, D. W. C.: Clustered event-triggered consensus analysis: An impulsive framework. IEEE Transactions Industrial Electronics. **63**, 7133–7143 (2016)
31. Chen, W.-H., Jiang, Z. Y., Lu, X. M., Luo, S. X.: H_∞ synchronization for complex dynamical networks with coupling delays using distributed impulsive control. Nonlinear Analysis: Hybrid Systems. **17**, 111–127 (2015)
32. He, W. L., Chen, G. R., Han, Q. -L., Qian, F.: Network-based leader-following consensus of nonlinear multi-agent systems via distributed impulsive control. Information Sciences. **380**, 145–158 (2017)
33. Lu, J. Q., Wang, Z. D., Cao, J. D., Ho, D. W. C., Kurths, J.: Pinning impulsive stabilization of nonlinear dynamical networks with time-varying delay. International Journal of bifurcation and chaos. **22**, 1250176 (2012)
34. Zhang, W. B., Tang, Y., Miao, Q. Y., Fang, J. -A.: Synchronization of stochastic dynamical networks under impulsive control with time delays. IEEE Transactions on Neural Networks and Learning Systems. **25**, 1758–1768 (2014)
35. Yang, Z. C., Xu, D. Y.: Stability analysis and design of impulsive control systems with time delay. IEEE Transactions on Automatic Control. **52**, 1448–1454 (2007)
36. Lakshmikantham, V., Bainov, D., Simenonv, P.: Theory of impulsive different equations. World Scientific, Singapore (1989)
37. Hu, J. Q., Liang, J. L., Cao, J. D.: Synchronization of hybrid-coupled heterogeneous networks: pinning control and impulsive control schemes. Journal of the Franklin Institute. **351**, 2600–2622 (2014)

Chapter 6
Quasi-Synchronization of Parameter Mismatched CDNs with Multiple Impulsive Effects

6.1 Introduction

In the past two decades, the collective behaviors of the coupled systems or interconnected systems have attracted increasing interests of the worldwide scholars from different research fields, such as computer science, mathematics, physic science and so on. Actually, for the coupled systems, there may emerge different forms of collective behaviors. Undoubtedly, it is known to us that the synchronization phenomenon of all or part systems is of great concern. For the basic concept of synchronization, it denotes that adjust the steps of the interconnected systems in order to make their behaviors uniform [1–6]. From the first study on synchronization to now, a number forms of synchronization patterns have been discovered and investigated like the (complete) synchronization [7], the impulsive synchronization [8], the phase synchronization [9], the cluster synchronization [5], projective synchronization [10] and the lag synchronization [11]. For these synchronization patterns, the synchronization errors should be expected to goes to zero with time elapsing.

Distinct to those synchronization patterns mentioned above, the quasi synchronization or called quasi-synchronization for short, could be regarded as one special synchronization phenomenon because for the interconnected systems in a digital communication network, they could be forced to be synchronized within a synchronization error set in advance, that means the synchronization error may not verge to zero as time goes on, and sometimes, it even performs with fluctuations. In fact, the quasi-synchronization phenomenon is resulted from some inevitable factors in the systems themselves, the control input and the communication channels in a complex dynamical network like the mismatched parameters existed in different systems, the effectiveness of the designed controllers and the disturbances in the communication channels [12–17]. For example, in [12], the dissipativity phenomenon and the global quasi-synchronization for a class of delay coupled neural networks was dis-

Z. Tang et al., *Impulsive Synchronization of Complex Dynamical Networks*,
https://doi.org/10.1007/978-981-16-5383-4_6

cussed by considering the concept of Filippov solutions because of the discontinuous nonlinear activation functions in each neural network. In addition, in [15], the quasi-synchronization problem of a kind of parameters mismatched chaotic systems was studied in view of the stochastic stability theorems. Furthermore, the synchronization issue of two interconnected and periodically driven plasma oscillators with the rich dynamical transition to the quasi-synchronized state was investigated by Vincent in [17].

In fact, as we all know that only a small fraction of complex networks, either in nature world or in the artificial society, can achieve the final synchronization state by regulating the system parameters, including the coupling strength, the inner connected matrices and the outer coupling matrices [1, 18]. That is, most of the complex networks not only in the nature world but also in the artificial society are supposed to be synchronized should be applied with some outer control forces. Until now, a variety of control protocols such as feedback control [19], sliding mode control [20, 21], adaptive control [22], intermittent control [2], pinning control [23], tracking control [24] and impulsive control [25] have been presented in order to force the complex dynamical networks to be synchronizes in different forms. Comparing with the general continuous control methods mentioned before, the intermittent control and impulsive control, as typical representatives of the discontinuous control protocols, only supply the control signal at short time intervals or even only at certain time-triggered time instants, which efficiently increase the utilizing of the communication channels and effectively saves the required control costs.

As we all know that the impulse signal could supply instantaneous motion or instant power for a controlled system. In particular, for communication networks with impulsive control or impulsive disturbance, according to how large the impulsive strength effects are provided, the dynamical behaviors of the current node will be influenced by its neighborhood at different impulsive instants. In fact, impulsive control is regarded as one of the typical discontinuous control methods, and is now drawing a great deal of scholars' attention from different research areas [26–33]. For instance, the stabilization issue of a class of nonlinearly delayed dynamical networks was studied by Lu et al. in [30] by applying the pinning impulsive control. Later, He et al. investigated a kind of quasi-synchronization issues for the heterogeneous complex dynamic networks based on designing the impulsive controllers in [32], in which some problems like how to estimate the synchronization error boundary, how many nodes should be pinned and how to design the coupling strength have been discussed in detail. In additional, the exponential synchronization for the coupled Lur'e networks with impulsive disturbance was investigated by Zhao et al. in [34] by designing only a single impulsive controller, where different functions of the impulsive strength effects have been discussed respectively.

For impulsive control problems or impulsive disturbance issues in synchronization analysis of complex networks, the impulsive effect coefficient or the so-called impulsive effect strength is closely related to the final conclusion on synchronization criteria. Therefore, in previous related works studied the impulsive control or impulsive disturbance on complex networks, different kind functions of the impulsive effect coefficient have been deeply discussed. In particular, for achieving the synchronization of complex dynamical networks, the impulsive effect strength is always assumed to take values in different intervals, such as $(-2, 0)$ [26], $(-1, 1)$ [29], $(0, 1)$ [32, 33], or other similar limited value ranges of impulsive effects. Most of previous works have considered the impulsive effects where they are beneficial and efficient to the synchronization of the considered complex networks. However, for impulsive control or impulsive disturbance problems, due to the impulsive signal generated from the impulse generators always suffers from unpredicted fluctuations or disturbances, therefore the impulsive effect may occur some inevitable changes, that is, the impulsive effect either plays positive functions or negative roles. It further implies that the impulsive effect could not only accelerate the synchronization convergence velocity but also impede and even destroy the stability of the coupled networks. Therefore, a natural question comes to us while studying the impulsive control or impulsive disturbance issues for coupled networks, that is, what king of efficient controllers should be designed in order to make synchronize such complex networks with regard to distinct functions of the impulsive effects. A intuitionistic thought is that if the impulsive effect works positively to the final synchronization, the impulsive controller could be designed with considering this kind of impulsive effects. On the contrary, if the impulsive effect is negatively acting on the network systems, some other controls strategies, such as feedback control could be introduced within the controller to neutralize the disadvantageous impulsive effects.

To the authors' knowledge reach, until now, there only a few results related on the quasi-synchronization of a kind of coupled neural networks by applying the impulsive control protocol. Due to the theoretical values and the practical application potential of this issue, it stimulates us to proceed this work. Based on the above analysis, we mainly discuss the exponential and global quasi-synchronization problem of a kind of coupled neural networks with mismatched system parameters and time-varying delay by presenting an effective impulsive controller in this chapter. In view of the concept of average impulsive interval and the extended comparison principle for general impulsive delayed systems, sufficient criteria are derived for the quasi-synchronization of the coupled neural networks, respectively, according to distinct functions of the impulsive effects. Finally, three numerical examples with different impulsive effects and feedback control strengths are presented to show the validity of the theoretical analysis and the designed control protocol. The main characteristics of this chapter can be concluded as follows:

(1) Different form the control method discussed in [32], an effective impulsive controller is designed, where both of the positive impulsive effects and the negative impulsive effects are considered simultaneously. In other words, the extensive value ranges for the impulsive effects taking are considered at the same time no matter they are beneficial to the final synchronization on coupled neural networks or not.
(2) According to distinct functions of the impulsive effects in network synchronization, the exponential and global convergence velocities of the coupled neural networks are estimated. In additional, different from the exponential complete synchronization pattern, the quasi-synchronization errors are evaluated with respect

to the extended formula for variation of parameters and some useful mathematical techniques.

(3) The concept of average impulsive interval is skillfully introduced to evaluate the times of generated impulse signals on the impulsive sequence in the time interval instead of directly utilizing the upper bound or lower bound of all impulsive intervals, which effectively reduce the conservatism of the quasi-synchronization conditions.

The main outline of this chapter is arranged as follows. In Sect. 6.2, the coupled neural network models are firstly given and then some necessary preliminaries, such as definitions, lemmas and assumptions are listed next. In Sect. 6.3, the exponential and global quasi-synchronization of the coupled neural networks is discussed by designing the impulsive pinning controller. Three numerical simulations are presented to verify the control protocol and the theoretical results in Sect. 6.4. Finally, the conclusion is drawn in Sect. 6.5.

Notations. $\mathbb{R}^n$ denotes the n-dimensional Euclidean space. $\mathbb{R}^{n\times n}$ is the set of $n \times n$ real matrices. $*$ stands for the symmetrical part in a matrix. $\mathrm{diag}\{\cdots\}$ stands for a diagonal matrix. The symbol $\|\cdot\|$ stands for the Euclid norm of the matrix or the vector. $\lambda_{\max}(L)$ and $\lambda_{\min}(L)$ stand for the largest eigenvalue and the smallest eigenvalue of matrix L, respectively, $\max_{1\le i\le N}\{\cdot\}$ and $\min_{1\le i\le N}\{\cdot\}$ are the maximum value and the minimum value, respectively. A positive definite (semi-definite) matrix L is denoted as $L > 0 (L \ge 0)$. I stands for the identity matrix. Let $\mathcal{C}([-h, 0]; \mathbb{R})$ be the space of continuous functions mapping $[t_0 - h, t_0]$ into $\mathbb{R}^n$ with the norm defined by $\|\varphi\| = \max_{1\le i\le n}\{sup_{t_0-h\le s\le t_0}|\varphi_i(s)|\}$. $D^+u(t) = \overline{\lim}_{h\to 0^+}\frac{u(t+h)-u(t)}{h}$, $u(t_k^+) = \lim_{h\to 0^+} u(t_k + h)$, $u(t_k^-) = \lim_{h\to 0^-} u(t_k + h)$.

6.2 Model Description and Preliminaries

6.2.1 Problem Formulation

Consider an array of coupled delay neural networks which consists of N nonidentical subnetworks

$$\begin{aligned}\dot{y}_i(t) = &-C_i y_i(t) + A_i \bar{f}_1(y_i(t)) + B_i \bar{f}_2(y_i(t - h(t))) + \bar{I}_i \\ &+ \kappa \sum_{j=1}^{N} l_{ij}\Gamma y_j(t) + u_i(t),\end{aligned} \tag{6.1}$$

where $y_i(t) = [y_i^1(t), y_i^2(t), \ldots, y_i^n(t)]^T \in \mathbb{R}^n$ is the state vector of the i-th network at time t for $i = 1, 2, \ldots, N$; Matrix $C_i = \mathrm{diag}\{c_i^1, c_i^2, \ldots, c_i^n\}$ is diagonal positive definite with c_i^j denoting the rate with which the j-th neuron will reset its potential to the resting state when disconnected from the neural network and external input;

$A_i \in \mathbb{R}^{n\times n}$, $B_i \in \mathbb{R}^{n\times n}$ stand for the connection weight matrices; The constant $\kappa > 0$ denotes the coupling strength and $\Gamma = \text{diag}\{\gamma_1, \gamma_2, \ldots, \gamma_n\} \in \mathbb{R}^{n\times n}$ is called the inner-connected matrix with $\gamma_j \geq 0$ for $j = 1, 2, \ldots, n$; $\bar{I}_i = [\bar{I}_i^1, \bar{I}_i^2, \ldots, \bar{I}_i^n]^T \in \mathbb{R}^n$ is external input vector for $i = 1, 2, \ldots, N$; $\bar{f}_1, \bar{f}_2 : \mathbb{R}^n \to \mathbb{R}^n$ are two nonlinear activation functions, and they are continuous differentiable on $\mathbb{R}$ with $\bar{f}_1(0) = \bar{f}_2(0) = 0$; In particular, $\bar{f}_1(yt)) = [\bar{f}_1^1(y_i(t)), \bar{f}_1^2(y_i(t)), \ldots, \bar{f}_1^n(y_i(t))]^T$, $\bar{f}_2(y_i(t)) = [\bar{f}_2^1(y_i(t-h(t))), \bar{f}_2^2(y_i(t-h(t))), \ldots, \bar{f}_2^n(y_i(t-h(t)))]^T$; Time-varying function $h(t)$ is the delay satisfying $0 \leq h(t) \leq h$; Matrix $L = [l_{ij}]_{N\times N} \in \mathbb{R}^{N\times N}$ is the so-called outer coupling matrix which determined by the topology structure of the networks, furthermore, $l_{ij} > 0$ if there exists a connection from the i-th network to the j-th network $l_{ij} \neq l_{ji} (i \neq j)$, and $l_{ij} = 0$, otherwise; In addition, matrix L is assumed to satisfy the diffusive condition, that is, $l_{ii} = -\sum_{j\neq i j=1}^{N} l_{ij}$, and L is irreducible; The controller $u_i(t)$ will be designed in detail later.

Suppose that the coupled neural networks (6.1) satisfies the following initial values $y_i(t) = \psi_i(t) \in \mathcal{C}([-h, 0], \mathbb{R}^n)$ for $i = 1, 2, \ldots, N$ where $\mathcal{C}([-h, 0], \mathbb{R}^n)$ denotes the set of all continuous functions from $[-h, 0]$ to $\mathbb{R}^n$.

Let the error vector be $e_i(t) = y_i(t) - y_0(t)$ for $i = 1, 2, \ldots, N$, and $y_0(t)$ is the solution vector to the following isolated neural network

$$\dot{y_0}(t) = -Cy_0(t) + A\bar{f}_1(y_0(t)) + B\bar{f}_2(y_0(t-h(t))) + \bar{I}, \tag{6.2}$$

which contains the distinct system parameters from the coupled neural networks (6.1), and the vector is defined as $y_0(t) = [y_0^1(t), y_0^2(t), \ldots, y_0^n(t)]^T \in \mathbb{R}^n$, which could be thought as a chaotic orbit, a periodic orbit or even an equilibrium point.

In this chapter, suppose that any initial value $y_0(t)$ with $-h \leq t \leq 0$, there exist a time instant $T(0)$ and a positive constant $\bar{M}$ such that $\|y_0(t)\| \leq \bar{M}$ for all $t \geq T(0)$, which implies $y_0(t)$ is bounded with the upper boundary $\bar{M}$.

Considering the controlled coupled neural networks (6.1) and the target neural network (6.2) gives the following error neural networks

$$\begin{aligned}\dot{e}_i(t) = &-C_i e_i(t) + A_i f_1(e_i(t)) + B_i f_2(e_i(t-h(t))) \\ &+ \kappa \sum_{j=1}^{N} l_{ij}\Gamma e_j(t) + H_i(y_0(t)) + u_i(t),\end{aligned} \tag{6.3}$$

where nonlinear activation functions $f_1(e_i(t)) = \bar{f}_1(y_i(t)) - \bar{f}_1(y_0(t))$, $f_2(e_i(t-h(t))) = \bar{f}_2(y_i(t-h(t))) - \bar{f}_2(y_0(t-h(t)))$ and the difference $H_i(y_0(t)) = -(C_i - C)y_0(t) + (A_i - A)\bar{f}_1(y_0(t)) + (B_i - B)\bar{f}_2(y_0(t-h(t))) + (\bar{I}_i - \bar{I})$.

For realizing the quasi-synchronization between the coupled neural networks (6.1) and the target neural network (6.2), we present the following impulsive pinning controller

$$u_i(t) = -d_i e_i(t) + \sum_{k=1}^{\infty} \mu e_i(t)\delta(t - t_k), \tag{6.4}$$

where d_i is the nonnegative constant called the negative feedback control gain for $i = 1, 2, \ldots, N$ and $\mu \neq -1$ is the so-called impulsive effect which will be discussed in detain later, the impulse time series $\zeta = \{t_1, t_2, \ldots\}$ is a strictly increasing time sequence for each impulsive instant, it strictly satisfies the condition $t_{k-1} < t_k$ and it has $\lim_{k\to+\infty} t_k = +\infty$, furthermore, the function $\delta(\cdot)$ is the famous Dirac impulsive function.

With thinking the impulsive pinning controller (6.4) as designed before, we could rewrite the controlled coupled error neural networks as follows

$$\begin{cases} \dot{e}_i(t) = -(C_i + d_i I_n)e_i(t) + A_i f_1(e_i(t)) + B_i f_2(e_i(t - h(t))) \\ \qquad\qquad + \kappa \sum_{j=1}^{N} l_{ij}\Gamma e_j(t) + H_i(y_0(t)), \qquad t \neq t_k, \\ \Delta e_i(t_k) = e_i(t_k^+) - e_i(t_k^-) = \mu e_i(t_k^-), \; k = 1, 2, \ldots. \end{cases} \tag{6.5}$$

Denote $D = \text{diag}\{d_1, d_2, \ldots, d_n\}$ as the negative feedback control gain matrix.

Throughout this chapter, it is supposed that the error state vector $e_i(t)$ is right-hand continuous at each impulsive instant $t = t_k (k = 1, 2, \ldots)$. Thus, it could be concluded that the solutions of the error neural networks (6.5) are piecewise right-hand continuous functions with the discontinuous point at each impulsive instant $t = t_k$ for $k = 1, 2, \ldots$. Furthermore, define the coupled error neural networks (6.5) satisfies the following initial values $e_i(t) = \phi_i(t) \in \mathcal{C}([-h, 0], \mathbb{R}^n)$ for $i = 1, 2, \ldots, N$.

Remark 6.1 The synchronization issues studied in this chapter could be thought as a kind of leader-follower problems. In particular, the target neural network (6.2) could be regarded as the leader while the other nonidentically coupled neural networks in (6.1) could be viewed as its followers. Although it is difficult to achieve the complete (or called full) synchronization for coupled neural networks due to the mismatch phenomena existed in the system parameters, the almost synchronization could be realized within some given error boundary $\bar{e}$, this phenomenon is the so-called quasi-synchronization. Therefore, our purpose in this chapter is to present the conditions for achieving the quasi-synchronization between the coupled neural networks (6.1) and the target neural network (6.2), and give the precise estimation on the synchronization error $\bar{e}$ by some mathematic technics, simultaneously.

6.2.2 *Preliminaries*

In this subsection, some useful definitions, lemmas and assumptions needed in this chapter for analysis the main results are presented firstly.

Definition 6.1 ([35]) Consider the nonidentically coupled neural networks (6.1) and the target neural network (6.2). For any initial conditions $\phi_i(t) \in \mathcal{C}([-h, 0], \mathbb{R}^n)$, if there exists a compact set $\bar{M}$ such that the error vector $e_i(t)$ in the controlled coupled error neural networks (6.5) converges into the ball $\bar{M} = \{e(t) \in \mathbb{R}^{nN} | \|e(t)\| \leq \bar{e}\}$ as $t \to +\infty$, then the quasi-synchronization among the coupled neural networks (6.1) and the target neural network (6.2) is eventually realized with an given error upper boundary $\bar{e} > 0$, where vector $e(t) = [e_1^T(t), e_2^T(t), \ldots, e_N^T(t)]^T$ is the compact form of each error state vector.

Definition 6.2 ([36]) For the impulsive sequence $\zeta = \{t_1, t_2, \ldots\}$ in the time interval (t, T), let the symbol $N_\zeta(T, t)$ be the number of impulsive times on the impulsive sequence ζ. If there exist two nonnegative constants N_0 and T_a satisfying

$$\frac{T - t}{T_a} - N_0 \leq N_\zeta(T, t) \leq \frac{T - t}{T_a} + N_0, \ \forall T \geq t \geq 0, \tag{6.6}$$

then it gives that the average impulsive interval of the above impulsive sequence ζ will be no larger than T_a.

Lemma 6.1 ([37]) *Considering the irreducible and non-negative off-diagonal matrix L, assume that it satisfies the zero-row-sum condition, then there hold the following conclusions*

(1) If $\delta \neq 0$ is an eigenvalue of the above mentioned matrix L, then it gives $\Re(\delta) < 0$, where $\Re()$ denotes the real part of the eigenvalue.
(2) For the above mentioned matrix L, there exists the eigenvalue 0 with multiplicity 1. Furthermore, the right eigenvectors corresponding to the eigenvalue 0 could be expressed as $k[1, 1, \ldots, 1]^T$ for any $k \neq 0$.
(3) Suppose that the normalized left eigenvector of matrix L corresponding to the eigenvalue 0 is presented as $\xi = [\xi_1, \xi_2, \ldots, \xi_N]^T \in \mathbb{R}^N$, and satisfies with $\max_{1 \leq i \leq N}\{\xi_i\} \leq 1$, then $\xi_i > 0$ for all $i = 1, 2, \ldots, N$.

Lemma 6.2 ([38]) *Let $\phi(t)$ is a continuous function except at some finite number of points t_k where $\phi(t_k^+) = \phi(t_k)$ and $\phi(t_k^-)$ exist. Define the set $PC(s) = \{\phi | \phi : [-h, \infty) \to \mathbb{R}^s\}$ and $0 \leq h(t) \leq h$. Consider two functions $U(t)$ and $V(t)$ from the set $PC(s)$ with setting $s = 1$. Suppose that there exist two constants α, β and σ satisfying*

$$\begin{cases} D^+u(t) \leq \alpha u(t) + \beta u(t - h(t)), \ t \neq t_k \\ u(t_k) \leq \sigma u(t_k^-), \ k = 1, 2, \ldots, \end{cases}$$

and

$$\begin{cases} D^+v(t) > \alpha v(t) + \beta v(t - h(t)), \ t \neq t_k \\ v(t_k) = \sigma v(t_k^-), \ k = 1, 2, \ldots. \end{cases}$$

If $U(t) \leq V(t)$ for $-h \leq t \leq 0$, then it further gives the conclusion that $U(t) \leq V(t)$ for $t > 0$.

Assumption 6.1 Suppose that two nonlinear activation functions $\bar{f}_1(\cdot) = [\bar{f}_1^1(\cdot), \bar{f}_1^2(\cdot), \ldots, \bar{f}_1^n(\cdot)]^T$, $\bar{f}_2(\cdot) = [\bar{f}_2^1(\cdot), \bar{f}_2^2(\cdot), \ldots, \bar{f}_2^n(\cdot)]^T$ satisfy the Lipschitz condition, that is, there exist two nonnegative constants $\bar{q}_{ij}$ and $\tilde{q}_{ij}$ satisfying

$$|\bar{f}_1^i(x) - \bar{f}_1^i(z)| \leq \sum_{j=1}^{n} \bar{q}_{ij}|x_j - z_j|,$$

$$|\bar{f}_2^i(x) - \bar{f}_2^i(z)| \leq \sum_{j=1}^{n} \tilde{q}_{ij}|x_j - z_j|,$$

for any $x = [x_1, x_2, \dots, x_n]^T \in \mathbb{R}^n$, $z = [z_1, z_2, \dots, z_n]^T \in \mathbb{R}^n$ and $i, j = 1, 2, \dots, n$. Then, we denote matrices $\bar{Q} = [\bar{q}_{ij}]_{n\times n}$ and $\tilde{Q} = [\tilde{q}_{ij}]_{n\times n}$.

Remark 6.2 In the controlled coupled error neural networks (6.3), the nonlinear function $H_i(y_0(t))$ denotes the heterogeneities in the distinct neural networks due to the mismatch phenomenon of the system parameters among the coupled neural networks (6.1) and the target neural network (6.2). It should be noticed that the initial value of $y_0(t)$ is bounded with the upper boundary $\bar{M}$ for any $t \in [-h, 0]$. Consider that two nonlinear activation functions $\bar{f}_1(\cdot)$ $\bar{f}_2(\cdot)$ are Lipschitz functions, therefore, $H_i(y_0(t))$ is also bounded. That is, there must exist a nonnegative constant $\bar{h}_i$ satisfying $\sup_{t\geq\bar{T}} \|H_i(y_0(t))\| = \bar{h}_i$ where $\bar{T} > T(0)$.

6.3 Main Results

In this section, the quasi-synchronization among the controlled coupled neural networks (6.1) and the target neural network (6.2) with parameter mismatches and time-varying delay will be investigated by imposing the designed impulsive pinning controller (6.4).

Theorem 6.1 *Consider the controlled coupled error neural networks (6.5) satisfying Assumption 6.1. For the impulsive sequence $\zeta = \{t_1, t_2, \dots\}$ defined in Definition 6.2, assume that the average impulsive interval of is no larger than T_a. Define $\rho = (1+\mu)^2$. If there exist three general matrices $P > 0$, $W_{i1} > 0$, $W_{i2} > 0$, two diagonal matrices $\Lambda > 0$, $\Phi > 0$, some scalars $\alpha_i > 0$, $\beta_i > 0$, $l_i > 0$, $\bar{h}_i > 0$, $\alpha = \min_{1\leq i\leq N}\{\alpha_i\}$ and $\beta = \max_{1\leq i\leq N}\{\beta_i\}$ for $i = 1, 2, \dots, N$, satisfy with the following two different cases.*

***Case I**: For impulsive effect $-2 < \mu \leq 0$ and $\mu \neq -1$, if*
***(i)** Matrix*

$$\Pi_i = \begin{bmatrix} \Pi_{i11} & 0 & PA_i & PB_i & P \\ * & \tilde{Q}^T\Lambda\tilde{Q} - \beta_i P & 0 & 0 & 0 \\ * & * & -\Phi & 0 & 0 \\ * & * & * & -\Lambda & 0 \\ * & * & * & * & -W_{i1} + W_{i2} \end{bmatrix}$$

is a negative definite matrix, where $\Pi_{i11} = -P(C_i + d_i I_n) - (C_i + d_i I_n)^T P + \bar{Q}^T\Phi\bar{Q} + \alpha_i P$ for $i = 1, 2, \dots, N$;

(ii) *Matrices* $W_{i1} - W_{i2} - l_i I \leq 0$;
(iii-a) *The inequality* $\rho^{-N_0}\beta + (\frac{\ln\rho}{T_a} - \alpha) < 0$,
then the trajectory of the controlled coupled error neural networks (6.5) globally and exponentially converges into the compact set $\bar{M}$ *with the convergence velocity* $\frac{\lambda}{2}$, *where the compact set* $\bar{M}$ *is defined as*

$$\bar{M} = \Bigg\{ e(t) \in \mathbb{R}^{nN} \mid \|e(t)\| \leq e_0 = \sqrt{\frac{\rho^{-N_0}\sum_{i=1}^{N} l_i \bar{h}_i^2}{\lambda_{\min}(P)\cdot \min_{1\leq i\leq N}\{\xi_i\}\cdot(-\rho^{-N_0}\beta - (\frac{\ln\rho}{T_a} - \alpha))}} \Bigg\}, \tag{6.7}$$

and $\lambda > 0$ *is the unique solution to the defined parameter equation* $\lambda + (\frac{\ln\rho}{T_a} - \alpha) + \rho^{-N_0}\beta e^{\lambda h} = 0$. *That is, the quasi-synchronization between the coupled neural networks (6.1) and the target neural network (6.2) is finally realized with the upper error boundary* $\bar{e}$ *under the designed impulsive pinning controller (6.4).*

Case II*: For the impulsive effect satisfies* $\mu \leq -2$ *or* $\mu > 0$, *if both the conditions* ***(i)*** *and* ***(ii)*** *in* ***Case I*** *are all satisfied and besides*
(iii-b) *There exists the result* $\rho^{N_0}\beta + (\frac{\ln\rho}{T_a} - \alpha) < 0$,
then the trajectory of the controlled coupled error neural networks (6.5) globally and exponentially converges into the compact set $\bar{M}'$ *with the convergence velocity* $\frac{\lambda'}{2}$, *where the compact set* $\bar{M}'$ *is defined as*

$$\bar{M}' = \Bigg\{ e(t) \in \mathbb{R}^{nN} \mid \|e(t)\| \leq e_0 = \sqrt{\frac{\rho^{N_0}\sum_{i=1}^{N} l_i \bar{h}_i^2}{\lambda_{\min}(P)\cdot \min_{1\leq i\leq N}\{\xi_i\}\cdot(-\rho^{N_0}\beta - (\frac{\ln\rho}{T_a} - \alpha))}} \Bigg\}, \tag{6.8}$$

and $\lambda' > 0$ *is the unique solution to the defined parameter equation* $\lambda' + (\frac{\ln\rho}{T_a} - \alpha) + \rho^{N_0}\beta e^{\lambda' h} = 0$. *That is, the quasi-synchronization among the coupled neural networks (6.1) and the target neural network (6.2) is eventually realized with the upper error boundary* $\bar{e}$ *by applying the impulsive pinning controller (6.4).*

Proof Construct the Lyapunov functional as follows

$$V(e(t)) = \sum_{i=1}^{N} \xi_i e_i(t)^T P e_i(t), \tag{6.9}$$

where $\xi_i > 0$ is as defined in Lemma 6.1 satisfying $\max_{1\leq i\leq N}\{\xi_i\} \leq 1$. Denote $\Xi = \mathrm{diag}\{\xi_1, \xi_2, \ldots, \xi_N\} \in \mathbb{R}^{N\times N}$. Obviously, Ξ is a diagonal positive definite matrix.

Firstly, considering the impulsive instant $t = t_k (k = 1, 2, \ldots)$, we have the following results according to controlled coupled error neural networks (6.5)

$$\begin{aligned} V(e(t_k^+)) &= \sum_{i=1}^{N} \xi_i e_i^T(t_k^+) P e_i(t_k^+) \\ &= (1+\mu)^2 \sum_{i=1}^{N} \xi_i e_i^T(t_k^-) P e_i(t_k^-) \\ &= \rho V(e(t_k^-)). \end{aligned} \tag{6.10}$$

On the other hand, for the instant during the non-impulsive interval $t \in [t_{k-1}, t_k)$ $(k = 1, 2, \ldots)$, it gives

$$\begin{aligned} D^+V(e(t)) = 2\sum_{i=1}^{N} \xi_i e_i^T(t) P[-(C_i + d_i I_n) e_i(t) + A_i f_1(e_i(t)) \\ + B_i f_2(e_i(t - h(t))) + \kappa \sum_{j=1}^{N} l_{ij} \Gamma e_j(t) + H_i(y_0(t))] \\ = -2\sum_{i=1}^{N} \xi_i e_i^T(t) P(C_i + d_i I_n) e_i(t) + 2\sum_{i=1}^{N} \xi_i e_i^T(t) P A_i f_1(e_i(t)) \\ + 2\kappa \sum_{i=1}^{N}\sum_{j=1}^{N} \xi_i l_{ij} e_i^T(t) P \Gamma e_j(t) + 2\sum_{i=1}^{N} \xi_i e_i^T(t) P H_i(s) \\ + 2\sum_{i=1}^{N} \xi_i e_i^T(t) P B_i f_2(e_i(t - h(t))). \end{aligned} \tag{6.11}$$

With thinking that the coupling matrix L is an irreducible Laplacian matrix with non-negative off-diagonal elements, therefore, it gives

$$\lambda_{\max}(\Xi L + L^T \Xi) = 0.$$

Thus, by defining $\tilde{e}^k(t) = [e_1^k(t), e_2^k(t), \ldots, e_N^k(t)]^T$, it could further derive

$$\begin{aligned} &2\kappa \sum_{i=1}^{N}\sum_{j=1}^{N} \xi_i l_{ij} e_i^T(t) P \Gamma e_j(t) \\ &\le \kappa \lambda_{\max}(P) \sum_{k=1}^{n} \gamma_k \tilde{e}^k(t)^T (\Xi L + L^T \Xi) \tilde{e}^k(t) \\ &\le \kappa \lambda_{\max}(\Xi L + L^T \Xi) \lambda_{\max}(P) \sum_{k=1}^{n} \gamma_k \tilde{e}^k(t)^T \tilde{e}^k(t) = 0. \end{aligned}$$

Based on the above analysis, we could get the following results

$$
\begin{aligned}
D^+V(e(t)) \le & -2\sum_{i=1}^{N}\xi_i e_i^T(t)P(C_i+d_iI_n)e_i(t) + 2\sum_{i=1}^{N}\xi_i e_i^T(t)PA_i f_1(e_i(t)) \\
& + 2\sum_{i=1}^{N}\xi_i e_i^T(t)PH_i(y_0(t)) \\
& + 2\sum_{i=1}^{N}\xi_i e_i^T(t)PB_i f_2(e_i(t-h(t))).
\end{aligned} \tag{6.12}
$$

Regarding to Assumption 6.1, one could find two diagonal matrices $\Phi > 0$ and $\Lambda > 0$ satisfying that

$$
\begin{aligned}
& e_i^T(t)\bar{Q}^T\Phi\bar{Q}e_i(t) - f_1^T(e_i(t))\Phi f_1(e_i(t)) \ge 0, \\
& e_i^T(t-h(t))\tilde{Q}^T\Lambda\tilde{Q}e_i(t-h(t)) - f_2^T(e_i(t-h(t)))\Lambda f_2(e_i(t-h(t))) \ge 0.
\end{aligned} \tag{6.13}
$$

In view of the conclusions in (6.13), it further implies

$$
\begin{aligned}
D^+V(e(t)) \le & -2\sum_{i=1}^{N}\xi_i e_i^T(t)P(C_i+d_iI_n)e_i(t) + 2\sum_{i=1}^{N}\xi_i e_i^T(t)PA_i f_1(e_i(t)) \\
& + 2\sum_{i=1}^{N}\xi_i e_i^T(t)PH_i(y_0(t)) + 2\sum_{i=1}^{N}\xi_i e_i^T(t)PB_i f_2(e_i(t-h(t))) \\
& + \sum_{i=1}^{N}\xi_i e_i^T(t)\bar{Q}^T\Phi\bar{Q}e_i(t) - \sum_{i=1}^{N}\xi_i f_1^T(e_i(t))\Phi f_1(e_i(t)) \\
& + \sum_{i=1}^{N}\xi_i e_i^T(t-h(t))\tilde{Q}^T\Lambda\tilde{Q}e_i(t-h(t)) \\
& - \sum_{i=1}^{N}\xi_i f_2^T(e_i(t-h(t)))\Lambda f_2(e_i(t-h(t))) \qquad \le \sum_{i=1}^{N}\xi_i E_i^T(t)\bar{\Pi}_i E_i(t) + \sum_{i=1}^{N} H_i^T(y_0(t))W_{i1}H_i(y_0(t)),
\end{aligned}
$$

where the matrix is defined as follows

$$
\bar{\Pi}_i = \begin{bmatrix} \bar{\Pi}_{i11} & 0 & PA_i & PB_i & P \\ * & \tilde{Q}^T\Lambda\tilde{Q} & 0 & 0 & 0 \\ * & * & -\Phi & 0 & 0 \\ * & * & * & -\Lambda & 0 \\ * & * & * & * & -W_{i1} \end{bmatrix},
$$

and $\bar{\Pi}_{i11} = -P(C_i+d_iI_n) - (C_i+d_iI_n)^TP + \bar{Q}^T\Phi\bar{Q}$, the vector is defined as $E_i(t) = [e_i^T(t), e_i^T(t-h(t)), f_1^T(e_i(t)), f_2^T(e_i(t-h(t))), H_i^T(y_0(t))]^T$.

Considering two conditions **(i)** and **(ii)** in **Case I** of Theorem 6.1 gives

$$
\begin{aligned}
D^+V(e(t)) \le & -\sum_{i=1}^{N} \xi_i\alpha_i e_i^T(t)Pe_i(t) + \sum_{i=1}^{N} \xi_i\beta_i e_i^T(t-h(t))Pe_i(t-h(t)) \\
& + \sum_{i=1}^{N} H_i^T(y_0(t))(W_{i1}-W_{i2})H_i(y_0(t)) \\
\le & -\min_{1\le i\le N}\{\alpha_i\}\sum_{i=1}^{N} \xi_i e_i^T(t)Pe_i(t) + \sum_{i=1}^{N} l_i H_i^T(y_0(t))H_i(y_0(t)) \\
& + \max_{1\le i\le N}\{\beta_i\}\sum_{i=1}^{N} \xi_i e_i^T(t-h(t))Pe_i(t-h(t)) \\
\le & -\alpha\sum_{i=1}^{N} \xi_i e_i^T(t)Pe_i(t) + \sum_{i=1}^{N} l_i\bar{h}_i^2 + \beta\sum_{i=1}^{N} \xi_i e_i^T(t-h(t))Pe_i(t-h(t)) \\
= & -\alpha V(e(t)) + \beta V(e(t-h(t))) + \sum_{i=1}^{N} l_i\bar{h}_i^2,
\end{aligned} \tag{6.14}
$$

where we denote the parameters α and β as $\alpha = \min_{1\le i\le N}\{\alpha_i\}$ and $\beta = \max_{1\le i\le N}\{\beta_i\}$.

For any $\varepsilon > 0$, let $v(t)$ be the unique solution to the following impulsive system with time-varying delay

$$
\begin{cases}
\dot{v}(t) = -\alpha v(t) + \beta v(t-h(t)) + \sum_{i=1}^{N} l_i\bar{h}_i^2 + \varepsilon, \\
v(t_k^+) = \rho v(t_k^-), \quad k = 1, 2, \ldots, \\
v(t) = \lambda_{\max}(P)\sum_{i=1}^{N} \|\phi_i(t)\|^2, \quad -h \le t \le 0.
\end{cases} \tag{6.15}
$$

Therefore, system (6.15) could be thought as the comparison systems of $V(t)$ with respect to the results (6.10) and (6.14). Then, based on Lemma 6.2 in this chapter, it shows $V(t) \le v(t)$ for any $t > 0$.

Therefore, in view of the formula for the variation of parameters [39], one has the following integral equation for general state term $v(t)$ and time-varying delay term $v(t-h(t))$

$$
v(t) = C(t,0)v(0) + \int_0^t C(t,s)[\beta v(s-h(s)) + (\sum_{i=1}^{N} l_i\bar{h}_i^2 + \varepsilon)]ds, \ t \ge 0, \tag{6.16}
$$

where $C(t,s) (t \ge s \ge 0)$ is the Cauchy matrix for the following linear impulsive system

$$\begin{cases} \dot{v}(t) = -\alpha v(t), \ t \neq t_k, \\ v(t_k^+) = \rho v(t_k^-), \ k = 1, 2, \ldots. \end{cases}$$

In addition, the Cauchy matrix could be computed by the following formula

$$C(t, s) = e^{-\alpha(t-s)} \prod_{s \leq t_k \leq t} \rho.$$

In the following, in order to derive the conditions for the quasi-synchronization of the coupled neural networks with parameter mismatches and time-varying delay, two different cases with respect to distinct functions of the impulsive effects will be discussed, respectively.

Case I. Impulsive effect $-2 < \mu \leq 0$ **but** $\mu \neq -1$.
According to the definition $\rho = (1 + \mu)^2$ in Theorem 6.1, then we have $0 < \rho \leq 1$. Besides, the average impulsive interval of the impulsive sequence $\zeta = \{t_1, t_2, \ldots\}$ is assumed to be less than T_a, therefore, it follows from Definition 6.2 with $N_\zeta(t, s) \geq \frac{t-s}{T_a} - N_0$ that

$$\begin{aligned} C(t, s) &= e^{-\alpha(t-s)} \prod_{s \leq t_k \leq t} \rho \\ &\leq e^{-\alpha(t-s)} \rho^{N_\zeta(t,s)} \leq e^{-\alpha(t-s)} \rho^{(\frac{t-s}{T_a} - N_0)} \\ &\leq \rho^{-N_0} e^{\alpha(t-s)} e^{\frac{\ln \rho}{T_a}(t-s)} \leq \rho^{-N_0} e^{(\frac{\ln \rho}{T_a} - \alpha)(t-s)}. \end{aligned} \tag{6.17}$$

Substituting inequality of Cauchy matrix (6.17) into (6.16) yields

$$\begin{aligned} v(t) &= C(t, 0)v(0) + \int_0^t C(t, s)[\beta v(s - h(s)) + (\sum_{i=1}^N l_i \bar{h}_i^2 + \varepsilon)]ds \\ &\leq \rho^{-N_0} \lambda_{\max}(P) \sum_{i=1}^N \|\phi_i(0)\|^2 e^{(\frac{\ln \rho}{T_a} - \alpha)t} \\ &\quad + \int_0^t \rho^{-N_0} \cdot e^{(\frac{\ln \rho}{T_a} - \alpha)(t-s)} [\beta v(s - h(s)) + (\sum_{i=1}^N l_i \bar{h}_i^2 + \varepsilon)]ds \\ &\leq \sigma e^{(\frac{\ln \rho}{T_a} - \alpha)t} + \int_0^t e^{(\frac{\ln \rho}{T_a} - \alpha)(t-s)} [\rho^{-N_0} \beta v(s - h(s)) \\ &\quad + \rho^{-N_0} (\sum_{i=1}^N l_i \bar{h}_i^2 + \varepsilon)]ds, \end{aligned} \tag{6.18}$$

where the symbol $\sigma = \rho^{-N_0} \lambda_{\max}(P) \sup_{-h \leq s \leq 0} \sum_{i=1}^N \|\phi_i(s)\|^2 > 0$. Define the parameter function $\varpi(\lambda) = \lambda + (\frac{\ln \rho}{T_a} - \alpha) + \rho^{-N_0} \beta e^{\lambda h}$. It is easy to find that $\varpi(\lambda)$ is continuous for all λ. On the other hand, it gives that $\varpi(0) = (\frac{\ln\rho}{T_a} - \alpha) + \rho^{-N_0}\beta < 0$, $\varpi(\infty) > 0$ and $\dot{\varpi}(\lambda) = 1 + h\rho^{-N_0}\beta e^{\lambda h} > 0$. It further implies that $\varpi(\lambda)$ is monotone about the variable λ. Therefore, the defined parameter function $\varpi(\lambda) = 0$ has an unique solution $\lambda > 0$.

Due to $0 < \rho \leq 1$, $\varepsilon > 0$, $\lambda > 0$, it obtains the following result for $-h \leq t \leq 0$

$$\begin{aligned} v(t) &\leq \rho^{-N_0}\lambda_{\max}(P)\sum_{i=1}^{N}\|\phi_i(t)\|^2 \\ &< \sigma e^{-\lambda t} + \frac{\rho^{-N_0}(\sum_{i=1}^{N} l_i\bar{h}_i^2 + \varepsilon)}{-(\frac{\ln\rho}{T_a} - \alpha) - \rho^{-N_0}\beta}. \end{aligned} \tag{6.19}$$

In the following, the results in the above inequality (6.19) will be proved for all $t > 0$. In particular, there holds

$$v(t) < \sigma e^{-\lambda t} + \frac{\rho^{-N_0}(\sum_{i=1}^{N} l_i\bar{h}_i^2 + \varepsilon)}{-(\frac{\ln\rho}{T_a} - \alpha) - \rho^{-N_0}\beta}, \quad \forall t > 0. \tag{6.20}$$

For this purpose, we shall verify the result in inequality (6.20) in view of an useful mathematical method: proof by contradiction. Particularly, suppose that if (6.20) is no longer satisfied for all $t > 0$, as a consequence, there at least exists one instant $t^* > 0$ satisfying

$$v(t^*) \geq \sigma e^{-\lambda t^*} + \frac{\rho^{-N_0}(\sum_{i=1}^{N} l_i\bar{h}_i^2 + \varepsilon)}{-(\frac{\ln\rho}{T_a} - \alpha) - \rho^{-N_0}\beta}, \tag{6.21}$$

but for any $t < t^*$, it still satisfies

$$v(t) < \sigma e^{-\lambda t} + \frac{\rho^{-N_0}(\sum_{i=1}^{N} l_i\bar{h}_i^2 + \varepsilon)}{-(\frac{\ln\rho}{T_a} - \alpha) - \rho^{-N_0}\beta}, \quad t < t^*. \tag{6.22}$$

Letting the symbol $\frac{\ln\rho}{T_a} - \alpha = -c$ and Combining the equations $\varpi(\lambda) = 0$, (6.18) and (6.22), we have the following results

$$\begin{aligned} v(t^*) &\leq \sigma e^{(\frac{\ln\rho}{T_a} - \alpha)t^*} + \int_0^{t^*} e^{(\frac{\ln\rho}{T_a} - \alpha)(t^* - s)}[\rho^{-N_0}\beta v(s - h(s)) + \rho^{-N_0}(\sum_{i=1}^{N} l_i\bar{h}_i^2 + \varepsilon)]ds \\ &< e^{-ct^*}\{\sigma + \frac{\rho^{-N_0}(\sum_{i=1}^{N} l_i\bar{h}_i^2 + \varepsilon)}{c - \rho^{-N_0}\beta} \\ &\quad + \int_0^{t^*} e^{cs}[\rho^{-N_0}\beta v(s - h(s)) + \rho^{-N_0}(\sum_{i=1}^{N} l_i\bar{h}_i^2 + \varepsilon)]ds\} \end{aligned}$$

$$
\begin{aligned}
&< e^{-ct^*}\{\sigma + \frac{\rho^{-N_0}(\sum_{i=1}^N l_i\bar{h}_i^2+\varepsilon)}{c-\rho^{-N_0}\beta} + \int_0^{t^*} e^{cs}[\rho^{-N_0}\beta(\sigma e^{-\lambda(s-h(s))} \\
&\quad + \frac{\rho^{-N_0}(\sum_{i=1}^N l_i\bar{h}_i^2+\varepsilon)}{c-\rho^{-N_0}\beta}) + \rho^{-N_0}(\sum_{i=1}^N l_i\bar{h}_i^2+\varepsilon)]ds\} \\
&\le e^{-ct^*}\{\sigma + \frac{\rho^{-N_0}(\sum_{i=1}^N l_i\bar{h}_i^2+\varepsilon)}{c-\rho^{-N_0}\beta} + \frac{\rho^{-N_0}\beta\sigma}{c-\lambda}e^{\lambda h}[e^{(c-\lambda)t^*}-1] \\
&\quad + \frac{c\rho^{-N_0}(\sum_{i=1}^N l_i\bar{h}_i^2+\varepsilon)}{c-\rho^{-N_0}\beta}\frac{(e^{ct^*}-1)}{c}\} \\
&= \sigma e^{-ct^*} + \frac{\rho^{-N_0}(\sum_{i=1}^N l_i\bar{h}_i^2+\varepsilon)}{c-\rho^{-N_0}\beta}e^{-ct^*} + \frac{\rho^{-N_0}\beta\sigma}{c-\lambda}e^{\lambda h}[e^{(c-\lambda)t^*}-1]e^{-ct^*} \\
&\quad + \frac{\rho^{-N_0}(\sum_{i=1}^N l_i\bar{h}_i^2+\varepsilon)}{c-\rho^{-N_0}\beta}e^{-ct^*}(e^{ct^*}-1) \\
&= \sigma e^{-\lambda t^*} + \frac{\rho^{-N_0}(\sum_{i=1}^N l_i\bar{h}_i^2+\varepsilon)}{c-\rho^{-N_0}\beta}.
\end{aligned}
\tag{6.23}
$$

It is not difficult to find that the inequality (6.23) contradicts to the hypothesis (6.21), which implies the true of (6.20) for all $t > 0$. Let $\varepsilon \to 0$, it gives

$$
\lambda_{\min}(P)\min_{1\le i\le N}\{\xi_i\}\|e(t)\|^2 \le V(e(t)) \le v(t) < \sigma e^{-\lambda t} + \frac{\rho^{-N_0}\sum_{i=1}^N l_i\bar{h}_i^2}{-(\frac{\ln\rho}{T_a}-\alpha)-\rho^{-N_0}\beta},
$$

which further implies

$$
\begin{aligned}
\|e(t)\| \le \bar{e} = &\sqrt{\frac{\sigma}{\lambda_{\min}(P)\min_{1\le i\le N}\{\xi_i\}}}e^{-\frac{\lambda}{2}t} \\
&+ \sqrt{\frac{\rho^{-N_0}\sum_{i=1}^N l_i\bar{h}_i^2}{\lambda_{\min}(P)\cdot\min_{1\le i\le N}\{\xi_i\}[-(\frac{\ln\rho}{T_a}-\alpha)-\rho^{-N_0}\beta]}}.
\end{aligned}
\tag{6.24}
$$

Above all, as regard of the estimation on the error state $e(t)$ in (6.24), it could be detected that the compact set satisfies

$$
\bar{M} = \left\{ e(t)\in\mathbb{R}^{nN} \mid \|e\| \le \bar{e} = \sqrt{\frac{\rho^{-N_0}\sum_{i=1}^N l_i\bar{h}_i^2}{\lambda_{\min}(P)\cdot\min_{1\le i\le N}\{\xi_i\}[-(\frac{\ln\rho}{T_a}-\alpha)-\rho^{-N_0}\beta]}} \right\}
$$

as $t \to +\infty$.

The above discussion implies that the error neural network (6.5) globally and exponentially converges into the ball $\bar{M}$ at the convergence rate $\frac{\lambda}{2}$. That is, the quasi-synchronization among the coupled neural networks (6.1) and the target neural network (6.2) is eventually realized by giving the error $\bar{e}$. This completes the proof for the Case I.

Case II. Impulsive effect $\mu \leq -2$ **or** $\mu > 0$.
By the similar proof procedure with **Case I**, it holds the inequality $N_\zeta(t,s) \leq \frac{t-s}{T_a} + N_0$ based on the concept of average impulsive interval in Definition 6.2. Therefore, for the situation $\rho > 1$, i.e. $\mu \leq -2$ or $\mu > 0$, it obtains the following results

$$\begin{aligned} C(t,s) &= e^{-\alpha(t-s)} \prod_{s \leq t_k \leq t} \rho \\ &\leq e^{-\alpha(t-s)} \rho^{N_\zeta(t,s)} \leq e^{-\alpha(t-s)} \rho^{\frac{t-s}{T_a}+N_0} \\ &\leq \rho^{N_0} e^{-\alpha(t-s)} e^{\frac{\ln \rho}{T_a}(t-s)} \leq \rho^{N_0} e^{(\frac{\ln \rho}{T_a}-\alpha)(t-s)}. \end{aligned} \tag{6.25}$$

Substituting the estimation on Cauchy matrix (6.25) into (6.16) gives

$$\begin{aligned} v(t) &= C(t,0)v(0) + \int_0^t C(t,s)(\beta v(s-h(s)) + (\sum_{i=1}^N l_i \bar{h}_i^2 + \varepsilon))ds \\ &\leq \rho^{-N_0} \lambda_{\max}(P) \sum_{i=1}^N \|\phi_i(0)\|^2 e^{(\frac{\ln \rho}{T_a}+\alpha)t} \\ &+ \int_0^t \rho^{-N_0} e^{(\frac{\ln \rho}{T_a}+\alpha)(t-s)} [\beta u(s-h(s)) + (\sum_{i=1}^N l_i \bar{h}_i^2 + \varepsilon)]ds \\ &\leq \bar{\sigma} e^{(\frac{\ln \rho}{T_a}-\alpha)t} + \int_0^t e^{(\frac{\ln \rho}{T_a}-\alpha)(t-s)} [\rho^{N_0} \beta v(s-h(s)) \\ &+ \rho^{N_0} (\sum_{i=1}^N l_i \bar{h}_i^2 + \varepsilon)]ds. \end{aligned} \tag{6.26}$$

where the symbol $\bar{\sigma}$ is defined as $\bar{\sigma} = \rho^{N_0} \lambda_{\max}(P) \sup_{-h \leq s \leq 0} \sum_{i=1}^N \|\phi_i(s)\|^2 > 0$.

Similarly, it could be verified that the following conclusion is satisfied for all $t > 0$

$$v(t) < \bar{\sigma} e^{-\lambda t} + \frac{\rho^{N_0}(\sum_{i=1}^N l_i \bar{h}_i^2 + \varepsilon)}{-(\frac{\ln \rho}{T_a} - \alpha) - \rho^{N_0}\beta}, \quad \forall t > 0. \tag{6.27}$$

Letting $\varepsilon \to 0$ gives

$$\lambda_{\min}(P) \min_{1 \leq i \leq N} \{\xi_i\} \|e(t)\|^2 \leq V(e(t)) \leq v(t) < \bar{\sigma} e^{-\lambda t} + \frac{\rho^{N_0} \sum_{i=1}^N l_i \bar{h}_i^2}{-(\frac{\ln \rho}{T_a} - \alpha) - \rho^{N_0}\beta},$$

which further implies that

$$\|e(t)\| \le \bar{e} = \sqrt{\frac{\bar{\sigma}}{\lambda_{\min}(P)\min_{1\le i\le N}\{\xi_i\}}e^{-\frac{\lambda}{2}t}} + \sqrt{\frac{\rho^{N_0}\sum_{i=1}^{N} l_i \bar{h}_i^2}{\lambda_{\min}(P)\cdot\min_{1\le i\le N}\{\xi_i\}[-(\frac{\ln\rho}{T_a}-\alpha)-\rho^{N_0}\beta]}}. \tag{6.28}$$

By the same proof steps, it can be verified that the controlled coupled error neural network (6.5) globally and exponentially converges into the ball

$$\bar{M}' = \left\{ e \in \mathbb{R}^{nN} \mid \|e\| \le \bar{e} = \sqrt{\frac{\rho^{N_0}\sum_{i=1}^{N} l_i \bar{h}_i^2}{\lambda_{\min}(P)\cdot\min_{1\le i\le N}\{\xi_i\}[-(\frac{\ln\rho}{T_a}-\alpha)-\rho^{N_0}\beta]}} \right\}$$

with the convergence velocity $\frac{\lambda}{2}$. In the other words, the quasi-synchronization among the coupled neural networks (6.1) and the isolated target neural network (6.2) is eventually realized within the given error boundary $\bar{e}$ for the situation that $\rho > 1$.

Until now, we totally complete the proof. □

Remark 6.3 In this chapter, in order to realize the quasi-synchronization of the coupled neural networks, an effective impulsive pinning controller is designed, where the impulsive effect is supposed to satisfy with $\mu \neq -1$. Actually, during the proof steps, one could find that the impulsive effect μ is separately discussed by considering two possible value-taking ranges [36]: (a) $-2 < \mu \le 0$, $\mu \neq -1$; (b) $\mu \le -2$ or $\mu > 0$. It could be concluded that in each situation about the impulsive effect μ value-taking range, precisely synchronization error boundary and convergence velocity is presented. However, in some previous related works about impulsive control or impulsive disturbance like [25–29, 32] and references therein, the impulsive effect μ were always supposed to be satisfied with the conditions: (a) $0 < \mu < 1$ ([32]); (b) $-2 < \mu < 0$ ([26]); (c) $-1 < \mu < 1$ ([29]) or other similar restrictive value-taking ranges for μ. From the above conclusion, it could draw the conclusion that most of existed related works only focused on that the impulsive effect μ positively and effectively works to the network synchronization. However, more extensive value ranges for the impulsive effect μ taking should be deeply investigated in order to put the impulsive control schemes into wider practical applications. In this chapter, the value ranges for the impulsive effect taking are efficiently extended. No matter the impulsive effect plays a positive role or a negative role to the final stability of the coupled neural networks, the quasi-synchronization could be eventually realized accordingly by jointly applying the concept of average impulsive interval, the comparison principle on impulsive systems and the effective mathematical technical: proof by contradiction. Furthermore, it should be noticed that there are some special values should be further pointed out with considering the impulsive effect μ taking in this chapter. In the first place, if the impulsive effect $\mu = 0 (\rho = 1)$, i.e., $\Delta e_i(t_k) = 0$ or $e_i(t_k^+) = e_i(t_k^-)$, it shows that there is no impulse input effects on the coupled neural networks. For this situation, it could be thought as a special situation and it could be contained in Case I with $0 < \rho \leq 1$. Secondly, if the impulsive effect $\mu = -2(\rho = 1)$, that is $e_i(t_k^+) = -e_i(t_k^-)$, then it could be seem that the impulse inputs may impede the synchronization of the network systems. In this situation, the negative feedback control gain d_i could be set with some larger constants in order to counteract the negative effects brought by the negative impulse inputs. Thirdly, for the situation that the impulsive effect $\mu = -1(\rho = 0)$, it shows that $e_i(t_k) = e_i(t_k^+) = 0$, which is impossible in impulsive control issues, therefore, we excluded this situation in this chapter [30].

Remark 6.4 From the concept of average impulsive interval, it could be found that $N_\zeta(T, t)$ is the number of impulse times on the impulsive sequence ζ during the time interval (t, T). By regulating the nonnegative parameter N_0, the positive number T_a could be used to evaluate the number of impulse times during the time interval (t, T). Generally, there hold the inequalities $\min_k\{t_k - t_{k-1}\} \leq T_a \leq \max_k\{t_k - t_{k-1}\}$ for the positive number T_a. However, in some previous related results, such as [30, 31], the impulsive intervals were roughly set as $T_a = \min_k\{t_k - t_{k-1}\}$ or $T_a = \max_k\{t_k - t_{k-1}\}$. Therefore, by applying the concept of average impulsive interval, the conclusion derived in this chapter will be less conservative.

Remark 6.5 As can be found in the proof procedure, the Cauchy matrix $C(t, s)$ was firstly estimated in (6.17) and (6.25) for two different impulsive effects cases according to the concept of average impulsive interval, then the inequalities (6.18) and (6.26) were obtained correspondingly. Since the existence of the time-varying delay $v(s - h(s))$ in result (6.16), the general formula for the variation of parameters on linear impulsive system is no long applicable here as applied in [32, 33]. Therefore, by verifying the effectiveness of the results in inequality (6.19) for the situation $-h \leq t \leq 0$, the strict inequalities (6.20) and (6.27) were proved to be satisfied for all $t > 0$ with applying the mathematical method: proof by contradiction. According to the extended formula for the variation of parameters, the upper bound of the state $v(t)$ in comparison system (6.15) could be estimated as (6.20), where the convergence rate and the synchronization error also could be derived simultaneously.

Remark 6.6 As a matter of fact, the impulsive pinning controller (6.4) designed in this chapter includes two main function parts, i.e., the feedback control input and the impulsive control input. For the first situation $0 < \rho \leq 1$, that is, the impulsive effect μ plays a positive function to the synchronization of the coupled networks. Then the feedback control strength $d_i (i = 1, 2, \ldots, N)$ could be taken as zero for saving the control costs. While for the second situation $\rho > 1$, i.e., the impulsive effect μ could impede the synchronization of the coupled networks, result in the unstable or even destroy the coupled networks. For this situation, as a compensation, the feedback control term $-d_i e_i(t)$ could run efficiently to counteract the side-effects brought from the negative impulses. This point will be further illustrated by some numerical simulations in the next section.

Remark 6.7 In order to derive the upper bound of the vector in comparison system, there appears different methods. For instance, in view of the generalized Halanay inequalities for the Volterra functional differential equations, the upper bound of the synchronization error was evaluated by introducing some computations with p-norm on matrix [13]. Then the synchronization error was also obtained by estimating the selected Lyapunov function directly like Eq. (3.9) in [15], where the mathematical proof method: proof by contradiction was introduced. In addition, the synchronization error was acquired as well with respect to the minimum eigenvalue of matrix Π_1 in Eq. (64) by defining the functional $\tilde{V}(t, e_t) = e^{\sigma t} V(t, e_t)$, which is closely related to the Lyapunov-Krasovskii functional (10) in [16]. Different from these related methods, a delayed impulsive comparison system (6.15) was constructed in view of two inequalities (6.10) and (6.14) in this chapter according to Lemma 6.2. Then the extended formula for the variation of parameters and the concept of average impulsive interval for different functions of the impulsive effects μ were jointly applied in proof procedure to evaluate the upper bound of the vector $v(t)$ in the comparison system (6.15). As a result, the synchronization errors were calculated respect to different functions of impulsive effects based on the proof by contradiction methods. The difficulty in deriving the synchronization errors mainly lie in obtaining the upper bound of $v(t)$ by constructing the inequality (6.20) and (6.27).

As one of the special case for the main results in this chapter, we consider the identically coupled neural networks model (6.1). Then, the discussion would turn to be the complete synchronization between the following coupled neural networks

$$\begin{aligned} \dot{y}_i(t) = & -Cy_i(t) + A\bar{f}_1(y_i(t)) + B\bar{f}_2(y_i(t-h(t))) + \bar{I} \\ & + \kappa \sum_{j=1}^{N} l_{ij}\Gamma y_j(t), \quad i = 1, 2, \ldots, N \end{aligned} \tag{6.29}$$

and the target neural network (6.2). Correspondingly, we present the following theorem.

Theorem 6.2 *Consider the controlled identically coupled neural networks (6.5) with $H_i(y_0(t)) = 0$ $(i = 1, 2, \ldots, N)$. Suppose that Assumption 6.1 holds and the average impulsive interval of the impulsive sequence $\zeta = \{t_1, t_2, \ldots\}$ is no larger than T_a as defined in Definition 6.2. Let $\rho = (1+\mu)^2$. If there exist a matrix $P > 0$, two diagonal matrices $\Lambda > 0$, $\Phi > 0$, scalars $\alpha_i > 0$, $\beta_i > 0$ with $\alpha = \min_{1 \le i \le N}\{\alpha_i\}$ and $\beta = \max_{1 \le i \le N}\{\beta_i\}$ for $i = 1, 2, \ldots, N$ satisfying the following two different cases:*

***Case I**: For impulsive effect $-2 < \mu \le 0$ and $\mu \ne -1$, if matrices*

$$\Pi_i' = \begin{bmatrix} \Pi_{11}' & 0 & PA_i & PB_i \\ * & Q_2^T \Lambda Q_2 - \beta_i P & 0 & 0 \\ * & * & -\Phi & 0 \\ * & * & * & -\Lambda \end{bmatrix} \tag{6.30}$$

is negative definite, where $\Pi_{11}' = -P(C_i + d_i I) - (C_i + d_i I)^T P + Q_1^T \Phi Q_1 + \alpha_i P$, then the trajectory of the controlled coupled error neural networks (6.5) *with* $H_i(y_0(t)) = 0 (i = 1, 2, \ldots, N)$ *globally and exponentially converges to zero with the convergence velocity* $\frac{\lambda}{2}$, *where* $\lambda > 0$ *is the unique solution to the defined parameter equation* $\lambda + (\frac{\ln \rho}{T_a} - \alpha) + \rho^{-N_0}\beta e^{\lambda h} = 0$. *That is, the completely global and exponential synchronization between the identically coupled neural networks (6.29) and the target neural network (6.2) is finally realized under the designed impulsive pinning controller (6.4).*

Case II: *For impulsive effect* $\mu \leq -2$ *or* $\mu > 0$, *if condition (6.30) is satisfied, then the trajectory of the controlled coupled error neural networks (6.5) with* $H_i(y_0(t)) = 0 (i = 1, 2, \ldots, N)$ *globally and exponentially converges to zero with the convergence velocity* $\frac{\lambda'}{2}$, *where* $\lambda' > 0$ *is the unique solution to the defined parameter equation* $\lambda' + (\frac{\ln \rho}{T_a} - \alpha) + \rho^{N_0}\beta e^{\lambda' h} = 0$. *That is, the completely global and exponential synchronization between the identically coupled neural networks (6.29) and the target neural network (6.2) is finally realized under the designed impulsive pinning controller (6.4).*

Proof The proof procedure is similar to Theorem 6.1. We thus omit it here. □

6.4 Numerical Simulation

In this section, three numerical simulations will be present to verify the effectiveness of the main theorems in this chapter, where different functions of impulsive effects will be discussed by taking different values, respectively.

In this chapter, consider the neural networks with time-varying delay and different system parameters [40, 41] as follows

$$\dot{y}_i(t) = -C_i y_i(t) + A_i \bar{f}_1(y_i(t)) + B_i \bar{f}_2(y_i(t - h(t))) + \bar{I}_i, \quad i = 1, 2, 3, \tag{6.31}$$

$$y_i(t) = \begin{bmatrix} y_i^1(t) \\ y_i^2(t) \end{bmatrix}, \ \bar{f}_1(y_i(t)) = \begin{bmatrix} \bar{f}_1^1(y_i(t)) \\ \bar{f}_1^2(y_i(t)) \end{bmatrix}, \ \bar{f}_2(y_i(t - h(t))) = \begin{bmatrix} \bar{f}_2^1(y_i(t - h(t))) \\ \bar{f}_2^2(y_i(t - h(t))) \end{bmatrix},$$

$$C_1 = \begin{bmatrix} 1.0 & 0 \\ 0 & 1.0 \end{bmatrix}, \ A_1 = \begin{bmatrix} 2 & -0.1 \\ -5 & 3 \end{bmatrix}, \ B_1 = \begin{bmatrix} -1.5 & -0.1 \\ -0.2 & -2.5 \end{bmatrix},$$

$$C_2 = \begin{bmatrix} 0.8 & 0 \\ 0 & 1 \end{bmatrix}, \ A_2 = \begin{bmatrix} 2 & -0.11 \\ -5 & 3.2 \end{bmatrix}, \ B_2 = \begin{bmatrix} -1.6 & -0.1 \\ -0.18 & -2.5 \end{bmatrix},$$

$$C_3 = \begin{bmatrix} 0.9 & 0 \\ 0 & 1 \end{bmatrix}, \ A_3 = \begin{bmatrix} 2 & -0.11 \\ -5 & 3.1 \end{bmatrix}, \ B_3 = \begin{bmatrix} -1.6 & -0.1 \\ -0.18 & -2.6 \end{bmatrix},$$

and $\bar{I}_i = 0$ for $i = 1, 2, 3$. The target neural network could be described as

$$\dot{y}_0(t) = -C y_0(t) + A \bar{f}_1(y_0(t)) + B \bar{f}_2(y_0(t - h(t))) + \bar{I}, \tag{6.32}$$

$$y_0(t) = \begin{bmatrix} y_0^1(t) \\ y_0^2(t) \end{bmatrix}, \; C = \begin{bmatrix} 1.1 & 0 \\ 0 & 1 \end{bmatrix},$$

$$A = \begin{bmatrix} 2 & -0.09 \\ -5 & 3.1 \end{bmatrix}, \; B = \begin{bmatrix} -1.6 & -0.1 \\ -0.19 & -2.5 \end{bmatrix},$$

and $\bar{I} = 0$, $\bar{f}_1^j(u) = \bar{f}_2^j(u) = \tanh(u)$ with $u \in \mathbb{R}^2 (j = 1, 2)$ and time-varying delay is set as $h(t) = 1 + 0.3\sin(2t)$ which satisfies $h(t) \leq 1.3 = h$. In Fig. 6.1, we show

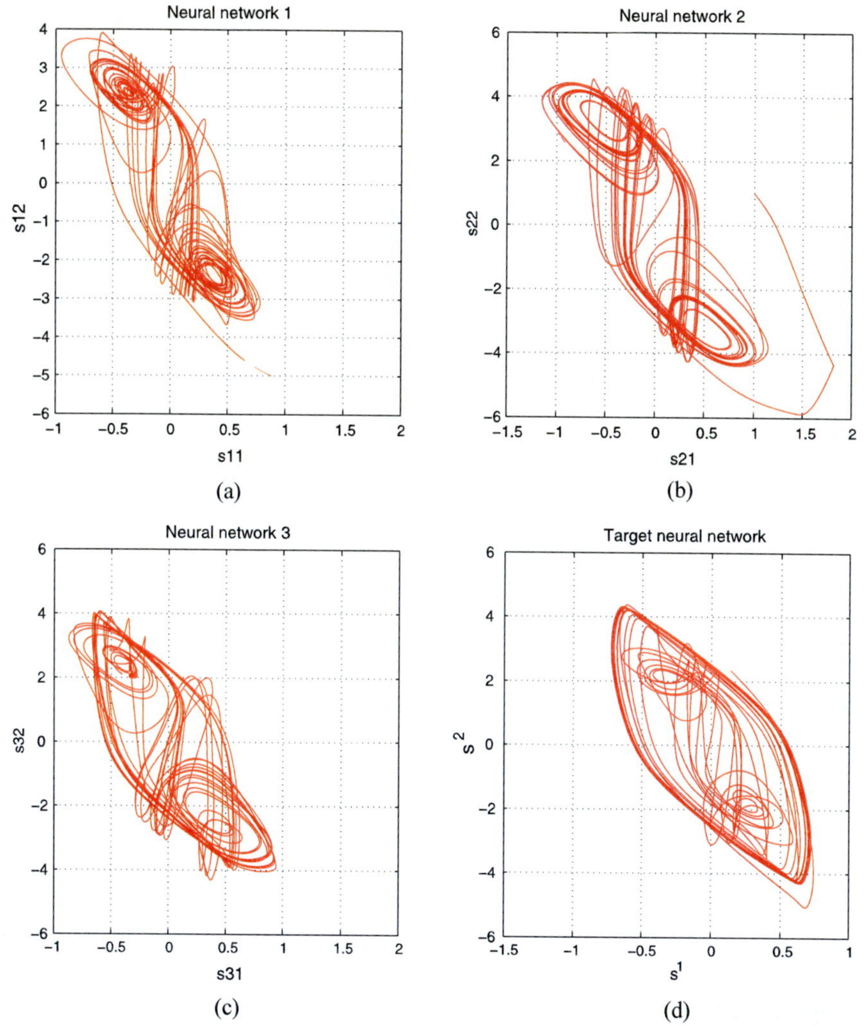

Fig. 6.1 **a–c** The phase graphs of three neural networks (6.31). **d** The target neural network (6.32)

the phase graphs of three neural networks described in (6.31) and the target neural network in (6.32).

Select the coupling matrix as $L = [1, -1, 0; -1, 2, -1; 0, -1, 1]$, the coupling strength as $c = 0.02$, the inner coupling matrix as $\Gamma = I_2$, the average impulsive interval is no larger than $T_a = 0.02$ and the positive constant $N_0 = 1$. Define the synchronization errors for each state as $E^j(t) = \sqrt{\frac{1}{N}\sum_{i=1}^{N}[e_i^j(t)]^2}$ for $j = 1, 2$.

Example 6.1 In the first example, set the negative feedback control strength as $d_i = 0$ $(i = 1, 2, \ldots, N)$, and select the impulsive effect $\mu = 0.2 > 0$. Then $\rho = (1 + \mu)^2 = 1.44 > 1$. Let the nonnegative parameters be $\alpha = 1672$ and $\beta = 326$, then it could be verified that $\rho^{N_0}\beta + (\frac{\log\rho}{T_a} - \alpha) = -1184.3 < 0$. Then the correspondingly error bound in Eq. (6.8) could be calculated as

$$\bar{e} = \sqrt{\frac{\rho^{N_0}\sum_{i=1}^{N} l_i \bar{h}_i^2}{\lambda_{\min}(P)\cdot\min_{1\le i\le N}\{\xi_i\}\cdot(-\Delta)}} = 0.0482.$$

From Theorem 6.1, it has been successfully verified that the quasi-synchronization between the nonidentically coupled neural networks (6.31) and the target neural network (6.32) could be realized globally and exponentially into the following compact set $\bar{M} = \{e \in \mathbb{R}^{nN} |\ \|e\| \le \bar{e} = 0.0482\}$.

In Fig. 6.2, pictures (a) and (b) depict two states evolution curves of the three coupled neural networks in (6.31) under the impulsive pinning controller with $\mu = 0.2$, while Fig. 6.3a shows the phase graphs of three neural networks. It can be concluded that the quasi-synchronization of the coupled neural networks is realized. In addition, one can find that the whole synchronization error, which is defined as $E(t) = \sqrt{\frac{1}{nN}\sum_{i=1}^{3}\sum_{j=1}^{2} e_i^j(t)^2}$, is under the bound 0.01963 as show in Fig. 6.3b, which is efficiently less than the error bound $\bar{e}$ calculated from the theorem. This fact further denotes that the quasi-synchronization between the coupled neural networks (6.31) with mismatched parameters and the target neural network (6.32) could be achieved within a given synchronization error bound.

Example 6.2 In the second example, set the impulsive effect as $\mu = -0.2$, i.e., $\rho = (1 - 0.2)^2 = 0.64 < 1$. With thinking the negative impulsive effects here, we set the negative feedback control strength as $d_i = 10$ $(i = 1, 2, \ldots, N)$. Similarly, there gives $\rho^{-N_0}\beta + (\frac{\log\rho}{T_a} - \alpha) = -1185.5 < 0$ and

$$\bar{e} = \sqrt{\frac{\rho^{-N_0}\sum_{i=1}^{N} l_i \bar{h}_i^2}{\lambda_{\min}(P)\cdot\min_{1\le i\le N}\{\xi_i\}\cdot(-\Delta')}} = 0.0496.$$

In Fig. 6.4, pictures (a) and (b) depict two states evolution curves of the three coupled neural networks in (6.31) under the impulsive pinning controller with $\mu = -0.2$, while Fig. 6.5a shows the phase graphs of three coupled neural networks. On the other hand, we verified from the Theorem 6.1 that the quasi-synchronization

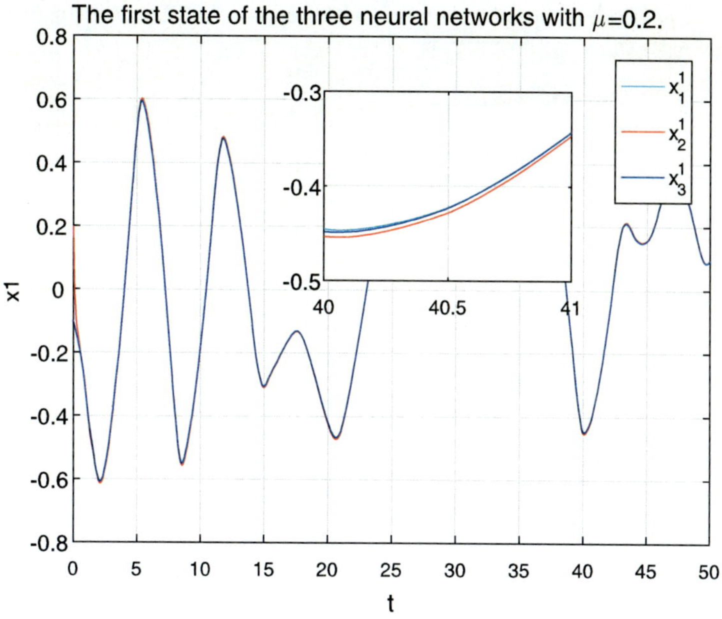

(a)

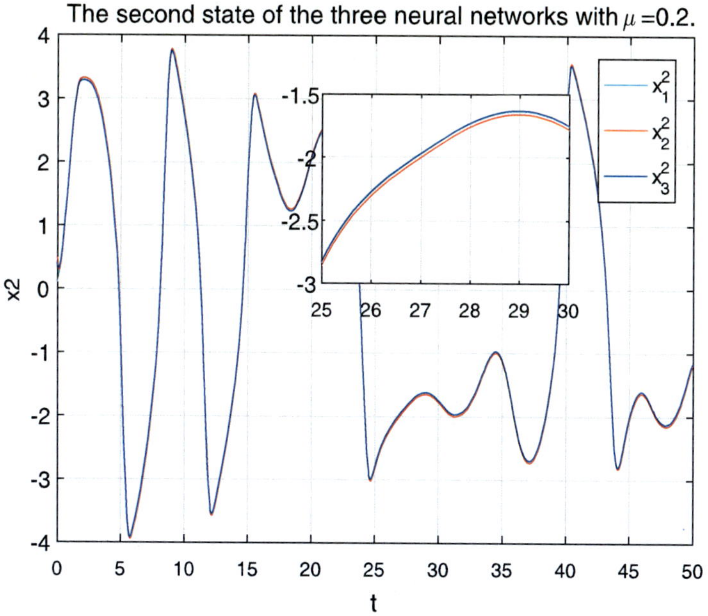

(b)

Fig. 6.2 **a**, **b** The error curves of two states with impulsive effect $\mu = 0.2$

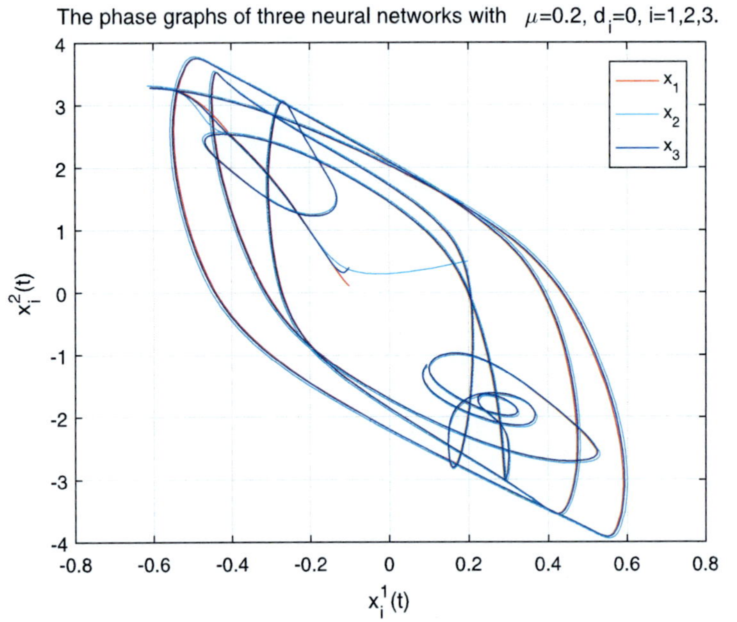

(a)

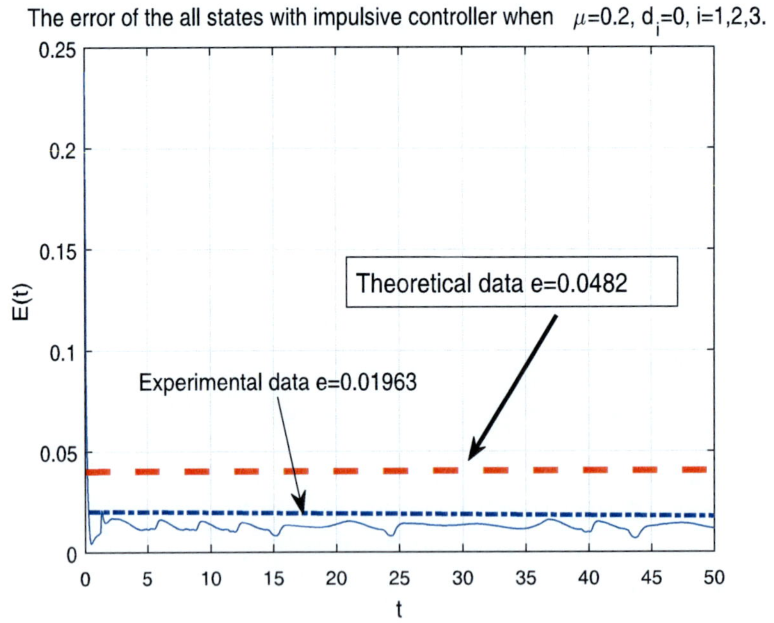

(b)

Fig. 6.3 **a** Phase graphs of three neural networks. **b** Final error curve analysis with $\mu = 0.2$

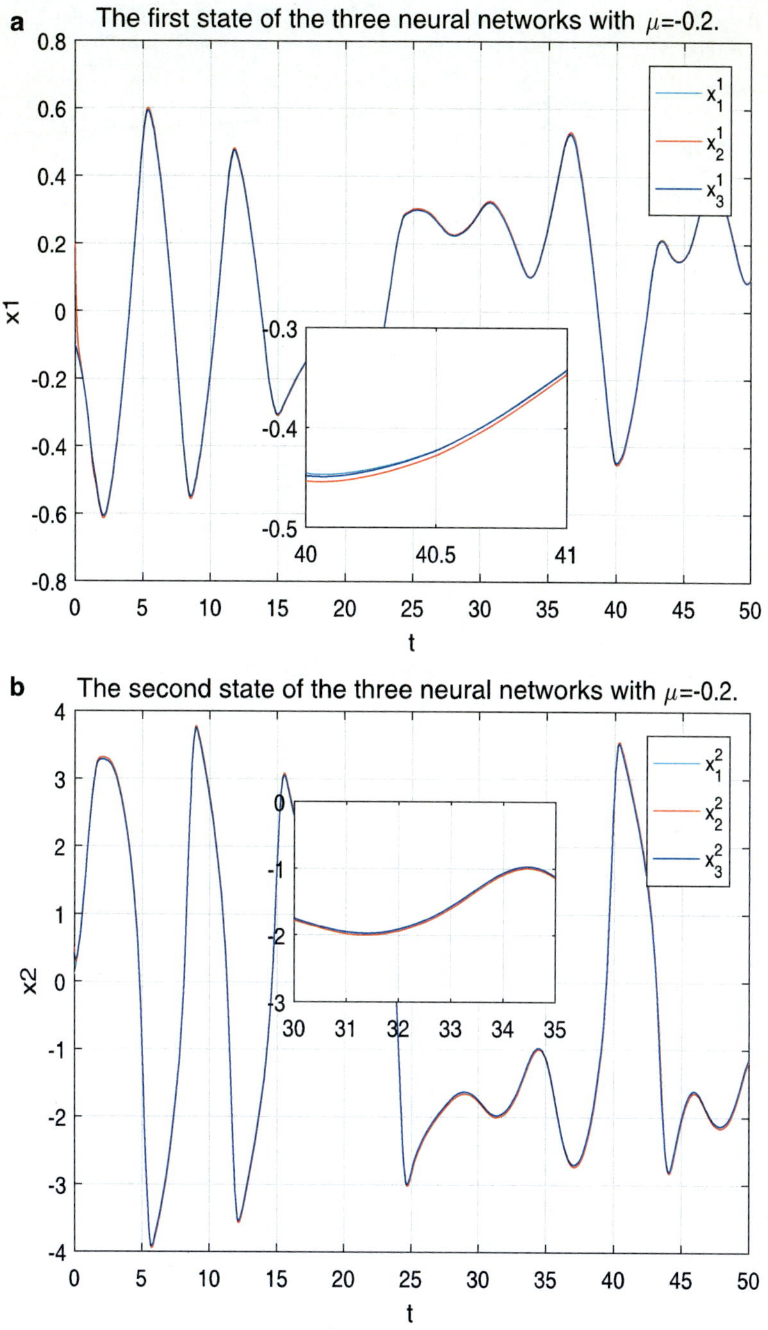

Fig. 6.4 **a**, **b** The error curves of two states with impulsive effect $\mu = -0.2$

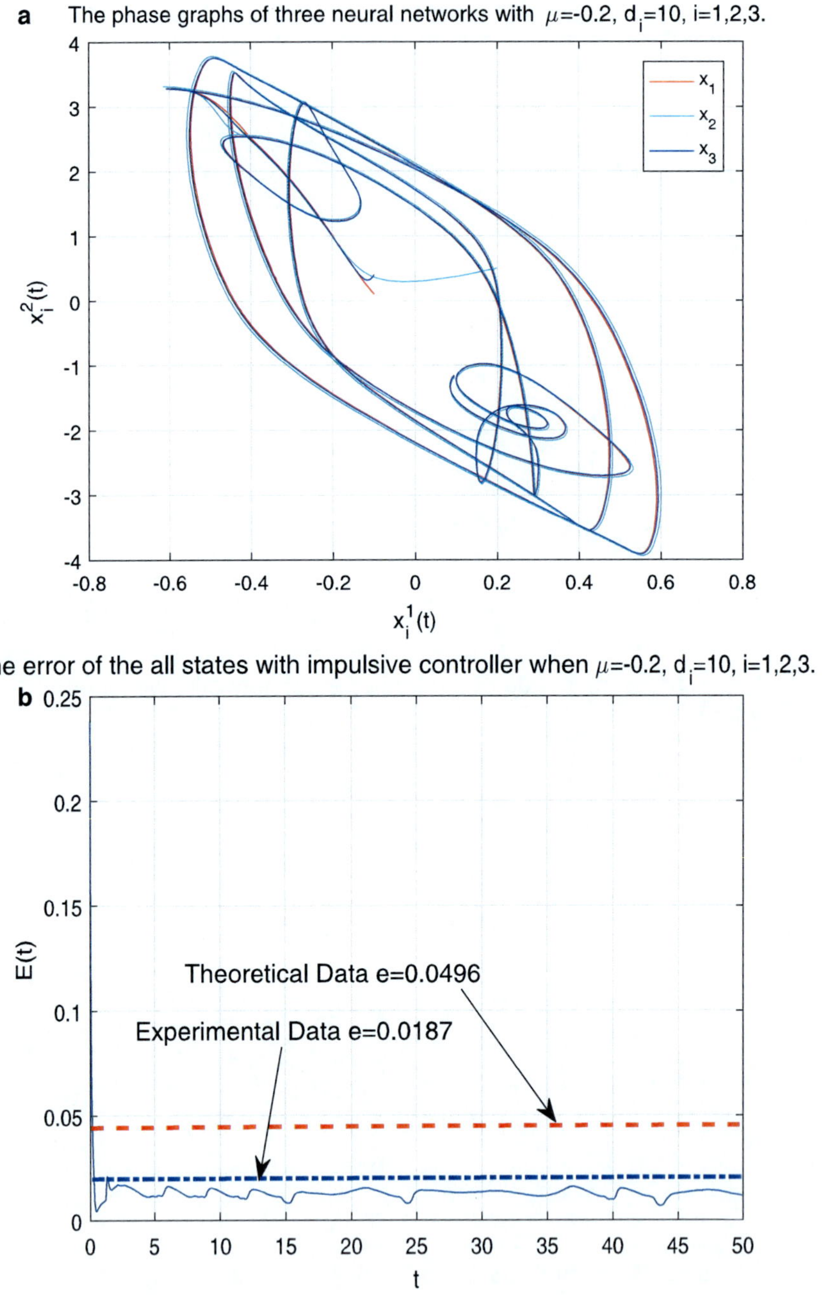

Fig. 6.5 **a** Phase graphs of three neural networks. **b** Final error curve analysis with $\mu = -0.2$, $d_1 = d_2 = d_3 = 1$

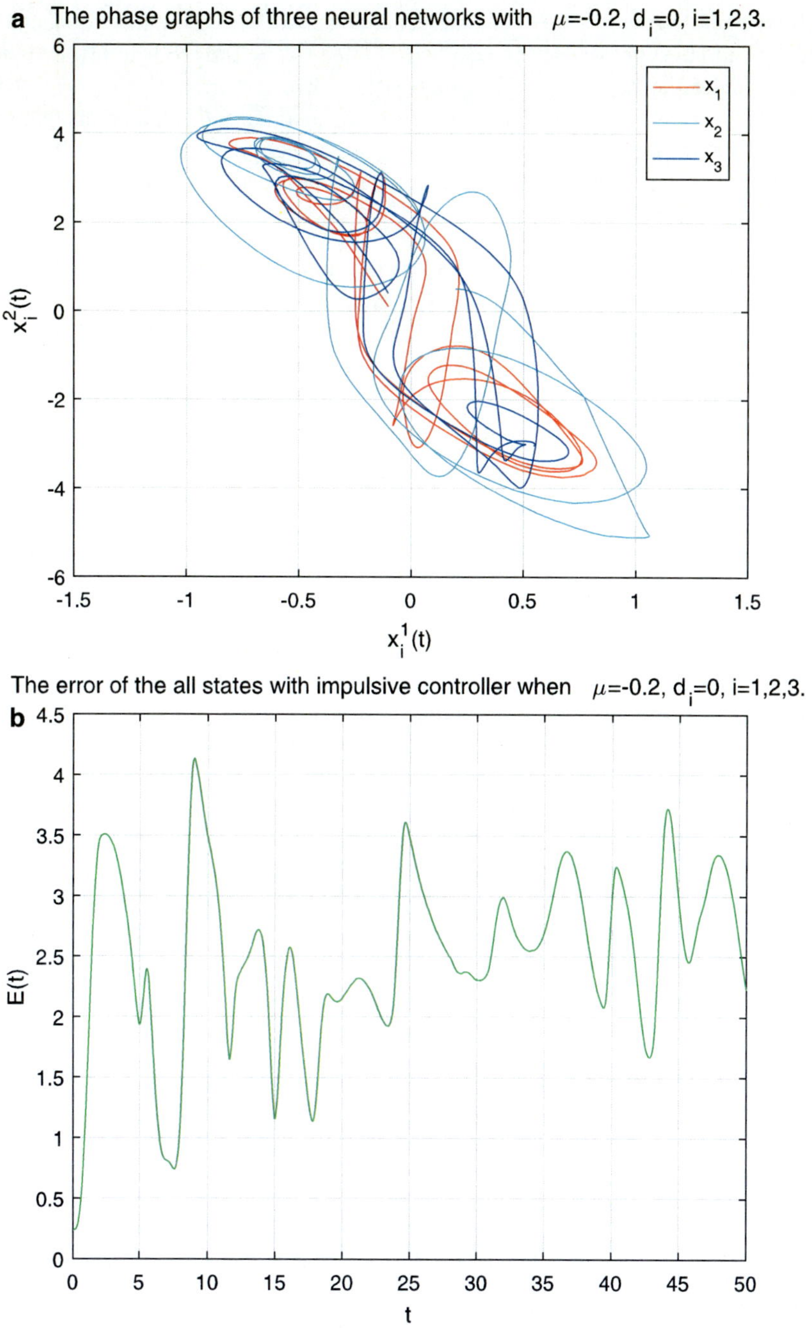

Fig. 6.6 **a** Phase graphs of three neural networks. **b** Final error curve analysis with $\mu = -0.2$, $d_1 = d_2 = d_3 = 0$

between neural networks (6.31) and target neural network (6.32) could exponentially achieved into the compact set $\bar{M} = \{e \in \mathbb{R}^{nN} | \ \|e\| \leq \bar{e} = 0.0496 \}$. In Fig. 6.5b, the error is under the bound 0.0187, which is less than the error bound $\bar{e} = 0.0496$ calculated from the theorem. Therefore, from the pictures and calculations, it could be found that the quasi-synchronization between the coupled neural networks (6.31) with mismatched parameters and the target neural network (6.32) could be achieved within the given error bound $\bar{e}$.

Example 6.3 In this example, also select the impulsive effect as $\mu = -0.2$, but set the negative feedback control strength as $d_i = 0$ $(i = 1, 2, 3)$. With the similar process, we plot the phase graphs and the error curve of the coupled neural networks in Fig. 6.6. It can be concluded that for the negative impulsive effect $\mu = -0.2$, the quasi-synchronization cannot be realized within a minor synchronization error. It further implies that, as we discussed in Remark 6.6, the feedback control term $-d_i e_i(t)$ could work as a compensation to counteract the side-effects brought from the negative impulses.

6.5 Conclusion

In this chapter, the impulsive pinning control for a kind of global and exponential quasi-synchronization problems on coupled neural networks with time-varying delay and parameter mismatches caused by the heterogeneity among neural networks have been discussed. Since the different functions of the impulsive effects in synchronization of the coupled networks, different values of impulsive effects for impulsive control protocols have been considered, respectively. Sufficient conditions have been derived for the quasi-synchronization on the nonidentically coupled neural networks in view of the extended comparison principle on impulsive systems, the concept of average impulsive interval and the extended formula for the variation of parameters. Furthermore, the global and exponential convergence velocities and the upper bounds of the synchronization error have been precisely estimated. In addition, three numerical simulations have been presented to verify the effectiveness of the main results proposed in this chapter.

References

1. Arenas, A., Guilera, A. D., Kurths, J., Moreno, Y., Zhou, C. S.: Synchronization in complex networks, *Physics Reports*, **469**, 93–153 (2008)
2. Liu, X. W., Chen, T. P.: Cluster synchronization in directed networks via intermittent pinning control, *IEEE Transactions on Neural Networks*, **22**, 1009–1020 (2011)
3. Feng, J. W., Tang, Z., Zhao, Y., Xu, C.: Cluster synchronization of nonlinearly coupled Lur'e networks with identical and non-identical nodes and an asymmetrical matrix, *IET Control Theory and Applications*, **7**, 2117–2127 (2013)

4. Tang, Y., Gao, H. J., Zou, W., Kurths, J.: Distributed synchronization in networks of agent systems with nonlinearities and ramdom switchings, *IEEE Transactions on Cybernetics*, **43**, 358–370 (2013)
5. Su, H. S., Rong, Z. H., Chen, M. Z. Q., Wang, X. F., Chen, G. R., Wang, H. W.: Decentralized adaptive pinning control for cluster synchronization of complex dynamical networks, *IEEE Transactions on Cybernetics*, **43**, 394–399 (2013)
6. Wu, Z. G., Shi, P., Su, H. Y., Chu, J.: Local synchronization of chaotic neural networks with sampled-data and saturating actuators, *IEEE Transactions on Cybernetics*, **44**, 2635–2645 (2014)
7. Yu, W. W., Chen, G. R., Lv, J. H., Kurths, J.: Synchronization via pinning control on general complex networks, *SIAM Journal of Control and Optimization*, **51**, 1395–1416 (2013)
8. Yang, X. S., Cao, J. D., Yang, Z. C.: Synchronization of coupled reaction-diffusion neural networks with time-varying delays via pinning-impulsive controller, *SIAM Journal of Control and Optimization*, **51**, 3486–3510 (2013)
9. Matheny, M. H., Grau, M., Villanueva, L. G., Karabalin, R. B., Cross, M. C., Roukes, M. L.: Phase synchronization of two anharmonic nanomechanical oscillators, *Physical Review Letters*, **112**, 394–399 (2014)
10. Du, H. Y., Shi, P., Lv, N.: Function projective synchronization in complex dynamical networks with time delay via hybrid feedback control, *Nonlinear Analysis: Real World Applications*, **14**, 1182–1190 (2013)
11. Zhang, Q. J., Zhao, J. H.: Projective and lag synchronization between general complex networks via impulsive control, *Nonlinear Dynamics*, **67**, 2519–2525 (2012)
12. Liu, X. Y., Chen, T. P., Cao, J. D., Lu, W. L.: Dissipativity and quasi-synchronization for neural networks with discontinuous activations and parameter mismatches, *Neural Networks*, **24**, 1013–1021 (2011)
13. He, W. L., Qian, F., Han, Q. -L., Cao, J. D.: Lag quasi-synchronization of coupled delayed systems with parameter mismatch, *IEEE Transactions on Circuits and Systems I: Regular Papers* , **58**, 1345–1357 (2011)
14. Pan, L. J., Cao, J. D.: Lag quasi-synchronization of coupled delayed systems with parameter mismatch, *Communications on Nonlinear Science and Numerical Simulation*, **17**, 1332–1343 (2012)
15. Liu, C., Li, C. D., Li, C. J.: Quasi-synchronization of delayed chaotic systems with parameters mismatch and stochastic perturbation, *Communications on Nonlinear Science and Numerical Simulation*, **16**, 4108–4119 (2011)
16. He, W. L., Qian, F., Han, Q. -L., Cao, J. D.: Synchronization error estimation and controller design for delayed Lur'e systems with parameter mismatches, *IEEE Transactions on Neural Networks and Learning Systems*, **23**,1551–1563 (2012)
17. Vincent, U. E., Kareem, S. O., Nbendjo, B. R. N., Njah, A. N.: Quasi-synchronization dynamics of coupled and driven plasma oscillators, *Chaos, Solitons and Fractals*, **70**, 85–94 (2015)
18. Li, Z. K., Duan, Z. S., Chen, G. R., Huang, L.: Consensus of multiagents systems and synchronization of complex networks: a unified viewpoint, *IEEE Transactions on Circuits and Systems-I: Regular Paper*, **57**, 213–224, (2010)
19. Wang, Y. W., Guan, Z. H., Wang, H. O.: Feedback and adaptive control for the synchronization of Chen system via a single variable, *Physical Letters A*, **312**, 34–40 (2003)
20. Wang, J. H., Gao, Y. B., Qiu, J. B., Ahn, C. K.: Sliding mode control for non-linear systems by Takagi–Sugeno fuzzy model and delta operator approaches, *IET Control Theory and Applications*, **11**, 1205–1213 (2017).
21. Liu, X. W., Chen, T. P.: Synchronization of nonlinear coupled networks via aperiodically intermittent pinning control, *IEEE Transactions on Neural Networks and Learning Systems*, **26**, 113–126 (2015)
22. Zhou, Q., Wu, C., Shi, P.: Observer-based Adaptive Fuzzy Tracking Control of Nonlinear Systems with Time Delay and Input Saturation, *Fuzzy Sets and Systems*, **316**, 49–68 (2017).
23. Wang, X. F., Chen, G. R.: Pinning control of scale-free dynamical network, *Physica A: Statistical Mechanics and its Applications*, **310**, 521–531 (2002)

24. Liang, H. J., Zhang, H. G., Wang, Z. S., Feng, T.: Reduced-order observer-based distributed tracking control for high-order multi-agent systems with heterogeneous leader, *Journal of the Franklin Institute*, **353**, 2511–2533 (2016)
25. Yang, X. S., Cao, J. D., Lu, J. Q.: Stochasitc synchronization of complex networks with nonidentical nodes via hybrid adaptive and impulsive control, *IEEE Transactions on Circuits and Systems-I: Regular Paper*, **59**, 371–384 (2012)
26. Lu, J. Q., Kruths, J., Cao, J. D., Mahdavi, N., Huang, C.: Synchronization control for nonlinear stochastic dynamical networks: pinning impulsive strategy, *IEEE Transactions on Neural Networks and Learning Systems*, **23**, 285–292 (2012)
27. Zhang, W. B., Tang, Y., Miao, Q. Y., Fang, J. -A.: Synchronization of stochastic dynamical networks under impulsive control with time delays, *IEEE Transactions on Neural Networks and Learning Systems*, **25**, 1758–1768 (2014)
28. Zhang, W. B., Tang, Y., Miao, Q. Y., Du, W.: Exponential synchronization of coupled switched neural networks with mode-dependent impulsive effects, *IEEE Transactions on Neural Networks and Learning Systems*, **24**, 1316–1326 (2013)
29. Lu, J. Q., Ho, D. W. C., Cao, J. D., Kurths, J.: Single impulsive controller for globally exponential synchronization of dynamical networks, *Nonlinear Analysis: Real World Applications*, **14**, 581–593 (2013)
30. Lu, J. Q,. Wang, Z. D., Cao, J. D., Ho, D. W. C., Kurths, J.: Pinning impulsive stabilization of nonlinear dynamical networks with time-varying delay, *International Journal of Bifurcation and Chaos*, **22**, 1250176:1–12 (2012)
31. Hu, J. Q., Liang, J. L., Cao, J. D.: Synchronization of hybrid-coupled heterogeneous networks: pinning control and impulsive control schemes, *Journal of the Franklin Institute*, **351**, 2600–2622 (2014)
32. He, W. L., Qian, F., Lam, J., Chen, G. R., Han, Q. -L., Kurths, J.: Quasi-synchronization of heterogeneous dynamic networks via distributed impulsive control: error estimation, optimization and design, *Automatica*, **62**, 249–262 (2015)
33. He, W. L., Qian, F., Cao, J. D., Han, Q. -L.: Impulsive synchronization of two nonidentical chaotic systems with time-varying delay, *Physical Letters A*, **375**, 498–504 (2011)
34. Zhao, Y., Tang, Z., Feng, J. W.: Single impulsive controller for exponential synchronization of stochstic Lur'e networks with impulsive distrubance, *Discrete Dynamics in Nature and Society*, **2013**, 864707:1–13 (2013)
35. Nishio, Y., Ushida, A.: Quasi-ynchronization phenomena in chaotic circuits coupled by one resistor, *IEEE Transactions on Circuits and Systems I: Fundamental Theory and Applications*, **43**, 491–496 (1996)
36. Lu, J. Q., Ho, D. W. C., Cao, J. D.: A unified synchronization criterion for impulsive dynamical networks, *Automatica*, **46**, 1215–1221 (2010)
37. Horn, R. A., Johnson, C. R.: *Matrix Analysis*, Cambridge University Press, New York, NY (1990)
38. Yang, Z. C., Xu, D. Y.: Stability analysis and design of impulsive control systems with time delay, *IEEE Transactions on Automatic Control*, **52**, 1448–1454 (2007)
39. Lakshmikantham, V., Bainov, D., Simenonv, P.: *Theory of impulsive different Equations*, World Scientific, Singapore (1989)
40. Wu, Z. G., Shi, P., Su, H. Y., Chu, J.: Stochastic synchronization of Markovian jump neural networks with time-varying delay using sampled data, *IEEE Transactions on Cybernetics*, **43**, 1796–1806 (2013)
41. Waibel, A., Hanazawa, T., Hinton, G., Shikano, K., Lang, K. J.: Phoneme recognition using time-delay neural networks, *IEEE Transactions on Acoustics, Speech, and Signal Processing*, **37**, 328–339 (1989)

Chapter 7
Cluster Synchronization of Nonlinearly Coupled Lur'e DNs: Impulsive Adaptive Control

7.1 Introduction

Recently, in view of the comprehensive engineering applications in real networks like biological networks and communication systems [1–4], the research on complex networks has attracted extensive attention. As the representative phenomenon of collective behaviors in complex networks, synchronization has become the hot issue of the investigation from scholars in different fields [5–8]. For example, in [9], the local synchronization for a kind of complex networks was discussed by Yu et al. In [10], Li et al. studied global synchronization of the delayed complex network. Besides, the impulsive synchronization [11] and cluster synchronization [12] are studied in depth.

There are some special cases in engineering applications, in which only the systems in the networks subgroup instead of the whole complex networks are required to reach synchronization. In this situation, the realization of complete synchronization may bring about the waste of control energy. Hence the thought of cluster synchronization is raised. In brief, cluster synchronization refers to the situation that all the nodes in each cluster will reach synchronization and there is no requirement for nodes among different clusters. In most previous available work studied about cluster synchronization [13, 14], only the cluster synchronization of identical complex networks with linearly coupled matrix is discussed. Especially, in [13], the cluster synchronization of linearly coupled complex networks is studied. However, assuming that the dynamics of the nodes in the complex networks are identical does not conform with the real world. For instance, communities in biological networks or in electrical networks.

In the latest few years, many control protocols have emerged. They can be compartmentalized into two parts according to time series. One is the continuous control protocols, such as adaptive control [15], pinning feedback control [16]. The other is discontinuous control methods, for example, intermittent control [17], impulsive

Z. Tang et al., *Impulsive Synchronization of Complex Dynamical Networks*,
https://doi.org/10.1007/978-981-16-5383-4_7

control [18, 19]. As the representative discontinuous control method, the impulsive control can reduce the control energy profoundly for it only works when the impulse generates. For example, in [18], with the designed impulsive controller, Tang et al. discussed the impulsive synchronization for derivative coupled neural networks. In [19], the problem of a class of stochastic reaction-diffusion dynamical networks was studied by presenting the comparison principle. By applying pinning control method, Wang et al. investigated the nonidentical networks with nonlinearly and asymmetrically coupled matrix [20]. Generally speaking, considering the effects of different external environments, almost all practical control systems are subjected to external disturbances, which may cause the signal delay. Apparently, this situation has not been taken into account in previous works [5, 21, 22], which may no fit the actual situation properly.

Inspired by the above works, the exponential cluster synchronization for the nonidentical Lur'e networks with asymmetrical matrices is deeply investigated in this chapter, in which the topic receives little concern by far. The distinguished parts of this chapter can be concluded as: (1) Different from previous works like [15], with the adaptive impulsive controller, the exponential cluster synchronization for the nonidentical Lur'e networks with nonlinearly coupled and asymmetrical matrix is discussed. (2) In [23], systems in the networks shared the same dynamics. While in this chapter, the adaptive pinning controllers are designed to imposed on those Lur'e systems which connect with other Lur'e systems of any other clusters, which is more practical in real environments. (3) By applying Lyapunov stability theorem, mathematical induction method and projection method, sufficient conditions for the exponential cluster synchronization of delayed Lur'e networks are gained in term of the average impulsive interval.

The main structure of this chapter is organized as follows. In Sect. 7.2, we mainly describe the Lur'e networks model and give some necessary definitions, lemmas and assumptions. In Sect. 7.3, the sufficient conditions will be presented to achieve exponential cluster synchronization. In Sect. 7.4, one numerical simulation example with impulsive effects is shown to verify the validity of the proposed results. Finally, conclusions are made in Sect. 7.5.

Notations. I_n denotes the identity matrix with n-dimension. $\mathbb{R}^{n\times n}$ is the set of $n \times n$ real matrices. $\mathbb{R}^n$ is the n-dimensional Euclidean space. The sign $\|\cdot\|$ stands for the Euclid norm of the vector or the matrix. diag$\{\cdots\}$ stands for a diagonal matrix. $\lambda_{\max}(Z)$ denotes the largest left eigenvalue of matrix Z, $\max_{1\leq i\leq n}\{\cdot\}$ and $\min_{1\leq i\leq n}\{\cdot\}$ are the maximum value and minimum value respectively. $D^+V(m) = \lim_{y\to 0^+} \frac{V(m+y)-V(m)}{y}$. $\mathfrak{N}^+$ denotes a set of positive integers. Normally, the matrices are supposed to meet the requirements of algebraic operation in the chapter.

7.2 Model Description and Preliminaries

7.2.1 Model Description

For the coupled Lur'e networks with time-varying delay and cluster-tree topology, there are some necessary assumptions should be defined. Consider the Lur'e networks with total m Lur'e systems and $\tilde{l}$ clusters with $m > \tilde{l} \geq 2$. Let $\mu_i = j$ if the i-th Lur'e system belongs to the j-th cluster. Denote U_j as the set of all Lur'e systems in the j-th cluster and the set $\bar{U}_j$ represents all Lur'e systems in the j-th cluster which connect with the Lur'e systems in any other clusters. Furthermore, we have (1). $U_i \cap U_j = \emptyset$, $i \neq j$ and $i,\ j = 1, 2, \ldots, \tilde{l}$; (2). $\cup_{i=1}^{\tilde{l}} U_i = \{1, 2, \ldots, m\}$.

Consider the following nonlinearly coupled Lur'e networks with $\tilde{l}$ clusters and time-varying delay

$$\begin{aligned}\dot{y}_i(t) = {} & B_{\mu_i} y_i(t) + A_{\mu_i} \bar{f}^1_{\mu_i}(H y_i(t)) + K_{\mu_i} \bar{f}^2_{\mu_i}(R y_i(t - \varrho_1(t))) \\ & + c \sum_{j=1}^{m} z_{ij} \Gamma G(y_j(t)) + u_i(t),\end{aligned} \tag{7.1}$$

where $y_i(t) = [y_i^1(t), y_i^2(t), \ldots, y_i^n(t)]^T \in \mathbb{R}^n$, $i = 1, 2, \ldots, m$ is the state variable of the i-th Lur'e system. $B_{\mu_i} \in \mathbb{R}^{n\times n}$, $A_{\mu_i} \in \mathbb{R}^{n\times l}$, $K_{\mu_i} \in \mathbb{R}^{n\times l}$, $H \in \mathbb{R}^{l\times n}$, $R \in \mathbb{R}^{l\times n}$ are constant matrices. The constant $c > 0$ denotes the coupling strength and $\Gamma = \text{diag}\{\gamma_1, \gamma_2, \ldots, \gamma_n\} \in \mathbb{R}^{n\times n}$ is the inner-linking matrix with $\gamma_i > 0$. In this chapter, for simple analysis, we assume that $\Gamma = I_n$; $\bar{f}^\pi_{\mu_i}(\cdot) : \mathbb{R}^l \to \mathbb{R}^l$ is memoryless nonlinear vector-valued function which is continuously differentiable on $\mathbb{R}$ for $\pi = 2$. $\varrho_1(t)$ stands for the time-varying delay which satisfies $0 \leq \varrho_1(t) \leq \bar{\varrho}$. $Z = (z_{ij})_{m\times m}$ represents the diffusive coupling matrix that shows the coupling configuration of the dynamical network. And we assume that matrix Z is considered to be asymmetric and irreducible in this chapter. Besides, $z_{ij} > 0$ if there are connections from the j-th Lur'e system to i-th Lur'e system for $i \neq j$, otherwise, $z_{ij} = 0$. Moreover, matrix Z satisfies $\sum_{j=1}^{m} z_{ij} = 0$. The nonlinear coupling function $G(\cdot) : \mathbb{R}^n \to \mathbb{R}^n$ is continuous with the form $G(y_i(t)) = [g_1(y_i^1(t)), g_2(y_i^2(t)), \ldots, g_n(y_i^n(t))]$. $u_i(t)$ is the control input of the Lur'e networks, which will be designed later.

Denote matrices $H = [h_1^T, h_2^T, \ldots, h_l^T]^T$ and $R = [r_1^T, r_2^T, \ldots, r_l^T]^T$ with $h_j, r_j \in \mathbb{R}^{1\times n}$ for $j = 1, 2, \ldots, l$. Then it derives $H y_i(t) = [h_1 y_i(t), h_2 y_i(t), \ldots, h_l y_i(t)]^T$, $\bar{f}^1_{\mu_i}(H y_i(t)) = [\bar{f}^{11}_{\mu_i}(h_1 y_i(t)), \bar{f}^{12}_{\mu_i}(h_2 y_i(t)), \ldots, \bar{f}^{1l}_{\mu_i}(h_l y_i(t))]^T$, and $R y_i(t - \varrho_1(t)) = [r_1 y_i(t - \varrho_1(t)), r_2 y_i(t - \varrho_1(t)), \ldots, r_l y_i(t - \varrho_1(t))]^T$, $\bar{f}^2_{\mu_i}(R y_i(t - \varrho_1(t))) = [\bar{f}^{21}_{\mu_i}(r_1 y_i(t - \varrho_1(t))), \bar{f}^{22}_{\mu_i}(r_2 y_i(t - \varrho_1(t))), \ldots, \bar{f}^{2l}_{\mu_i}(r_l y_i(t - \varrho_1(t)))]^T$, $i = 1, 2, \ldots, m$.

Consider the synchronization target as $\upsilon_{\mu_i}(t) = [\upsilon^1_{\mu_i}(t), \upsilon^2_{\mu_i}(t), \ldots, \upsilon^n_{\mu_i}(t)]^T \in \mathbb{R}^n$, $i = 1, 2, \ldots, \tilde{l}$.

$$\dot{\upsilon}_{\mu_i}(t) = B_{\mu_i}\upsilon_{\mu_i}(t) + A_{\mu_i}\bar{f}^1_{\mu_i}(H\upsilon_{\mu_i}(t)) + K_{\mu_i}\bar{f}^2_{\mu_i}(R\upsilon_{\mu_i}(t-\varrho_1(t))), \tag{7.2}$$

where $\lim_{t\to+\infty}\|\upsilon_{\mu_i} - \upsilon_{\mu_j}\| \neq 0$. The controller based on pinning control and impulsive control is elaborately designed as

$$u_i(t) = u_i^0(t) + u_i^1(t), \qquad i = 1, 2, \ldots, m, \tag{7.3}$$

in which the impulsive controller part is designed as

$$u_i^0(t) = \sum_{k=1}^{\infty}(\alpha\delta y_i(t) + \omega\delta y_i(t-\varrho_2(t)))\sigma(t-t_k), \tag{7.4}$$

and the part $u_i^1(t)$ based on pinning control scheme is designed as

$$u_i^1(t) = \begin{cases} -\epsilon_i(G(y_i(t)) - G(\upsilon_{\mu_i}(t))) - c\sum_{j=1}^m z_{ij}G(\upsilon_{\mu_j}(t)), \ i \in \bar{U}_{\mu_i}, \\ 0, \ i \in U_{\mu_i} \setminus \bar{U}_{\mu_i}, \end{cases} \tag{7.5}$$

where the constant $\epsilon_i \geq 0$ denotes the feedback control gain, α and ω are impulsive coefficients which relate to the error state and delayed error state, separately. $\sigma(\cdot)$ is the Dirac impulsive function and the time series $\zeta = \{t_1, t_2, \ldots\}$ is a strictly increasing impulsive instants sequence that satisfies $\lim_{k\to\infty} t_k = +\infty$ for $k \in \mathfrak{N}^+$. Define the control strength matrix $D = \mathrm{diag}\{\epsilon_1, \epsilon_2, \ldots, \epsilon_n\}$.

Remark 7.1 The part $-\epsilon_i(G(y_i(t)) - G(\upsilon_{\mu_i}(t)))$ in the controller (7.5) aims to realize the synchronization of all Lur'e systems in the same cluster, while the other item $-c\sum_{j=1}^m z_{ij}G(\upsilon_{\mu_j}(t))$ is designed to weaken the interconnection effects of Lur'e systems among different clusters.

Define the error vector as $\delta y_i(t) = y_i(t) - \upsilon_{\mu_i}(t)$. Therefor, we derive the following controlled error Lur'e networks

$$\begin{cases} \delta\dot{y}_i(t) = B_{\mu_i}\delta y_i(t) + A_{\mu_i}f^1_{\mu_i}(H\delta y_i(t)) + K_{\mu_i}f^2_{\mu_i}(R\delta y_i(t-\varrho_1(t))) \\ \qquad + c\sum_{j=1}^m z_{ij}(G(y_j(t)) - G(\upsilon_{\mu_j}(t))) \\ \qquad - \epsilon_i(G(y_i(t)) - G(\upsilon_{\mu_i}(t))), \quad t \neq t_k, \\ \Delta\delta y_i(t_k) = \alpha\delta y_i(t_k^-) + \omega\delta y_i(t-\varrho_2(t_k^-)), \quad t = t_k, \end{cases} \tag{7.6}$$

where $0 \leq \varrho_2(t) \leq \varrho$, $f^1_{\mu_i}(H\delta y_i(t)) = \bar{f}^1_{\mu_i}(Hy_i(t)) - \bar{f}^1_{\mu_i}(H\upsilon_{\mu_i}(t))$ and $f^2_{\mu_i}(R\delta y_i(t-\varrho_1(t))) = \bar{f}^2_{\mu_i}(Ry_i(t-\varrho_1(t))) - \bar{f}^2_{\mu_i}(R\upsilon_{\mu_i}(t-\varrho_1(t)))$. Throughout this chapter, $\delta y_i(t)$ is presumed to be right-hand continuous at $t = t_k$, $k \in \mathfrak{N}^+$, and $\delta y_i(t_k) = \delta y_i(t_k^+) = \lim_{t\to t_k^+}\delta y_i(t)$, $\delta y_i(t_k^-) = \lim_{t\to t_k^-}\delta y_i(t)$. Thus, we can get that the solutions of (7.6) are piecewise right-hand continuous functions with discontinuities at $t = t_k$ for $k = 1, 2, \ldots$.

7.2.2 Preliminaries

To derive the main results of this chapter, some preliminaries like definitions, lemmas and assumptions should be presented in advance.

Definition 7.1 ([24]) The Lur'e networks with time-varying delay (7.1) and the target Lur'e system (7.2) are said to achieve globally exponentially synchronization in mean square, if for any initial values $y_i(t_0) \in \mathbb{R}^n$, there exist constants $W_0, \lambda > 0$ and $t > t_0$ satisfying

$$\|y_i(t) - \upsilon(t)\| \leq W_0 \sup_{t_0 \in [t-\bar{\varrho}, t]} \|y_i(t_0) - \upsilon(t_0)\| e^{-\lambda(t-t_0)}.$$

Definition 7.2 ([21]) In the time interval (t, s), we consider an impulsive sequence $\zeta = \{t_1, t_2, \ldots\}$. Let $P_\zeta(s, t)$ be the impulsive times of the impulsive sequence ζ. It is considered that the average impulsive interval is less than $\hat{P}$ if there exist two positive numbers $\bar{P}$ and $\hat{P}$ satisfying

$$\frac{s-t}{\hat{P}} - \bar{P} \leq P_\zeta(s, t) \leq \frac{s-t}{\hat{P}} + \bar{P}, \quad \forall s \geq t \geq 0.$$

Definition 7.3 ([25]) Considering nonlinear function $g_k(\cdot) : \mathbb{R}^n \to \mathbb{R}^n$ belongs to the acceptable nonlinear coupling function class (NCF), that is, for arbitrary $x_1, x_2 \in \mathbb{R}$, the following inequality holds for $\varphi_k(\cdot) = g_k(\cdot) - \beta x$, where β and η are two non-negative scalars

$$| (g_k(x_1) - \beta x_1) - (g_k(x_2) - \beta x_2) | \leq \eta \mid x_1 - x_2 \mid .$$

Lemma 7.1 ([20]) *Let* $\boldsymbol{1}_n = \{1, 1, \ldots, 1\}^T \in \mathbb{R}^n$, $I_n = diag\{1, 1, \ldots, 1\} \in \mathbb{R}^{n\times n}$ *and* $\Phi = (\phi_{ij}) = I_n - \frac{1}{m}\boldsymbol{1}_n\boldsymbol{1}_n^T$. *For matrix* $Q \in \mathbb{R}^{n\times n}$ *satisfying* $q_{ii} = -\sum_{j=1, i\neq j}^{n} q_{ij}$, *there exists* $\varepsilon > 0$ *such that*

$$a^T Q b = a^T Q \Phi b \leq \frac{1}{2}(\frac{1}{\varepsilon} a^T Q Q^T a + \varepsilon b^T \Phi b).$$

Lemma 7.2 ([26]) *If matrix* $L = (l_{ij}) \in \mathbb{R}^{n\times n}$ *satisfies* $l_{ij} = l_{ji}$ *and* $l_{ii} = -\sum_{j=1, i\neq j}^{n} l_{ij}$, $i, j = 1, 2, \ldots, n$, *then the following equality holds for arbitrary two vectors* $m = (m_1, m_2, \ldots, m_n)^T$, $p = (p_1, p_2, \ldots, p_n)^T$

$$m^T L p = \sum_{i=1}^{n}\sum_{j=1}^{n} m_i l_{ij} p_j = -\sum_{j>i} l_{ij}(m_i - m_j)(p_i - p_j).$$

Lemma 7.3 ([27]) ι, ϑ *are positive parameters. Suppose a positive function* $V(t)$ *satisfies the scalar impulsive differential inequality*

$$\begin{cases} D^+V(t) \leq -\iota V(t) + \vartheta \sup_{\tau\in[t-\bar{\varrho},t]}\{V(\tau)\}, \\ V(t) = \Sigma(t), \qquad t \in [t_0 - \bar{\varrho}, t_0], \end{cases}$$

then one may obtain that $V(t) \leq \bar{V}(t_0)e^{-\lambda(t-t_0)}$, $t \geq t_0$, *where* $\Sigma(t)$ *is piecewise continuous and* $\bar{V}(t_0) = \sup_{\tau\in[t_0-\bar{\varrho},t_0]}\{V(\tau)\}$. *Besides,* $\lambda > 0$ *is the unique solution of the equation* $\lambda - \iota + \vartheta e^{\lambda\bar{\varrho}} = 0$.

Assumption 7.1 Assume the two nonlinear functions $\bar{f}^1_{\mu_i}(\cdot)$ and $\bar{f}^2_{\mu_i}(\cdot)$ satisfy the Lipschitz condition, i.e., there exist two positive constants p_{μ_i} and q_{μ_i}, for any $w, e \in \mathbb{R}^n$ and $i = 1, 2, \ldots, m$, the inequalities hold

$$\|\bar{f}^1_{\mu_i}(w) - \bar{f}^1_{\mu_i}(e)\| \leq p_{\mu_i}\|w - e\|,$$
$$\|\bar{f}^2_{\mu_i}(w) - \bar{f}^2_{\mu_i}(e)\| \leq q_{\mu_i}\|w - e\|.$$

7.3 Main Results

7.3.1 Synchronization for Lur'e Networks

In this subsection, the exponential cluster synchronization of the nonlinearly and non-identically coupled Lur'e networks (7.1) is investigated. Then, sufficient conditions will be derived by applying the impulsive pinning controllers (7.3).

Theorem 7.1 *Suppose that Assumption 7.1 holds, for the impulsive sequence* $\zeta = \{t_1, t_2, \ldots\}$, *suppose the average impulsive interval is no larger than* $\hat{P}$ *and* $g_k(\cdot) \in NCF(\beta, \eta)(k = 1, 2, \ldots, n)$ *for positive constants* $\beta > \eta$. *If there exist positive parameters* c, ν, ε *and the control strength matrix* D, *such that*
(i) *The matrix inequality*

$$M - (\beta - \eta)D + \nu I_m \leq 0;$$

(ii) *For the positive constant* μ, *it satisfies*

$$\{\Lambda + \Xi e^{\lambda\varrho}, e^{\lambda\bar{\varrho}}\} \leq \mu,$$

then the controlled error Lur'e networks (7.6) are globally and exponentially stable with the convergence rate $\frac{1}{2}(\lambda - \frac{\ln\mu}{\hat{P}})$, *where* $M = c\beta Z + \frac{c}{2\varepsilon}ZZ^T + c\eta^2\varepsilon(1 - \frac{1}{m})I_m$, $\Lambda = (1+\alpha)(1+\alpha+\omega)$, $\Xi = \omega(1+\alpha+\omega)$ *and* λ *is the unique feasible solution to the equation* $\lambda - \iota + \vartheta e^{\lambda\bar{\varrho}} = 0$ *with parameters* $\iota = -\min_{1\leq i\leq m}\{\lambda_{max}(B^s_{\mu_i} + A_{\mu_i}A^T_{\mu_i} + p^2_{\mu_i}H^TH + K_{\mu_i}K^T_{\mu_i} - 2\nu I_n)\}$ *and* $\vartheta = \max_{1\leq i\leq m}\{\lambda_{max}(q^2_{\mu_i}R^TR)\}$. *That is, the coupled Lur'e dynamical networks (7.1) is global and exponential synchronized to the target Lur'e system (7.2) by designing the impulsive pinning controllers (7.3).*

Proof Select Lyapunov function as:

$$V(t) = \frac{1}{2}\delta y(t)^T \delta y(t),$$

where $\delta y(t) = [\delta y_1(t), \delta y_2(t), \ldots, \delta y_m(t)]^T$.

Denote

$$\tilde{\delta} y^k(t) = [\delta y_1^k(t), \delta y_2^k(t), \ldots, \delta y_m^k(t)]^T,$$

$$\tilde{g}_k(y^k(t)) = [g_k(y_1^k(t)), g_k(y_2^k(t)), \ldots, g_k(y_m^k(t))]^T,$$

$$\tilde{g}_k(\upsilon^k(t)) = [g_k(\upsilon_{\mu_1}^k(t)), g_k(\upsilon_{\mu_2}^k(t)), \ldots, g_k(\upsilon_{\mu_m}^k(t))]^T,$$

$$\tilde{\varphi}_k(y^k(t)) = [\varphi_k(y_1^k(t)), \varphi_k(y_2^k(t)), \ldots, \varphi_k(y_m^k(t))]^T,$$

$$\tilde{\varphi}_k(\upsilon^k(t)) = [\varphi_k(\upsilon_{\mu_1}^k(t)), \varphi_k(\upsilon_{\mu_2}^k(t)), \ldots, \varphi_k(\upsilon_{\mu_m}^k(t))]^T$$

for $k = 1, 2 \ldots, n$.

First, at the impulsive instant $t = t_k$, $k \in \mathfrak{N}^+$, according to the error Lur'e networks (7.6), it derives

$$\begin{aligned}
V(t_k^+) &= \frac{1}{2}\delta y(t_k^+)^T \delta y(t_k^+) \\
&= \frac{1}{2}((1+\alpha)\delta y(t_k^-) + \omega \delta y(t_k^- - \varrho_2(t_k^-)))^T \\
&\quad \times ((1+\alpha)\delta y(t_k^-) + \omega \delta y(t_k^- - \varrho_2(t_k^-))) \\
&\le \frac{1}{2}(1+\alpha)(1+\alpha+\omega)\delta y(t_k^-)^T \delta y(t_k^-) \\
&\quad + \frac{1}{2}\omega(1+\alpha+\omega)\delta y(t_k^- - \varrho_2(t_k^-)))^T \delta y(t_k^- - \varrho_2(t_k^-)) \\
&= \Lambda V(t_k^-) + \Xi V(t_k^- - \varrho_2(t_k^-)).
\end{aligned} \tag{7.7}$$

Second, for $t \in [t_{k-1}, t_k)$, $k \in \mathfrak{N}^+$, calculating $D^+V(t)$ along the controlled error Lur'e networks (7.6) under controller (7.3) derives

$$\begin{aligned}
D^+V(t) &= \sum_{i=1}^{m} \delta y_i(t)^T \dot{\delta} y_i(t) \\
&= \sum_{i=1}^{m} \delta y_i(t)^T (B_{\mu_i} + A_{\mu_i} f_{\mu_i}^1(H\delta y_i(t)) + K_{\mu_i} f_{\mu_i}^2(R \\
&\quad \times \delta y_i(t - \varrho_1(t))) + c\sum_{j=1}^{m} z_{ij}(G(y_j(t)) - G(\upsilon_{\mu_j}(t))) \\
&\quad - \epsilon_i(G(y_i(t)) - G(\upsilon_{\mu_i}(t))))
\end{aligned}$$

$$
\begin{aligned}
&\leq \frac{1}{2}\sum_{i=1}^{m}\delta y_i(t)^T(2B_{\mu_i}+A_{\mu_i}A_{\mu_i}^T+p_{\mu_i}^2H^TH+K_{\mu_i}K_{\mu_i}^T)\delta y_i(t)\\
&\quad+\frac{1}{2}\sum_{i=1}^{m}q_{\mu_i}^2\delta y_i(t-\varrho_1(t))^TR^TR\delta y_i(t-\varrho_1(t)))\\
&\quad+c\sum_{i=1}^{m}\sum_{j=1}^{m}z_{ij}\delta y_i(t)^T(G(y_j(t))-G(\upsilon_{\mu_j}(t)))\\
&\quad-\sum_{i=1}^{m}\epsilon_i\delta y_i(t)^T(G(y_i(t))-G(\upsilon_{\mu_i}(t)))).
\end{aligned}
\tag{7.8}
$$

In addition, in terms of the coupling function $g_k(\cdot)\in NCF(\beta,\eta)$, one obtains

$$
\begin{aligned}
&c\sum_{i=1}^{m}\sum_{j=1}^{m}\delta y_i(t)^Tz_{ij}(G(y_j(t))-G(\upsilon_{\mu_j}(t)))\\
&=c\beta\sum_{k=1}^{n}\tilde{\delta}y^k(t)^TZ\tilde{\delta}y^k(t)\\
&\quad+c\sum_{k=1}^{n}\tilde{\delta}y^k(t)^TZ(\tilde{\varphi}_k(y^k(t))-\tilde{\varphi}_k(\upsilon^k(t))).
\end{aligned}
\tag{7.9}
$$

Focusing on the zero-row-sum matrix Z, on the light of Lemmas 7.1 and 7.2, there exists a positive parameter ε such that

$$
\begin{aligned}
&c\sum_{k=1}^{n}\tilde{\delta}y^k(t)^TZ(\tilde{\varphi}_k(y^k(t))-\tilde{\varphi}_k(\upsilon^k(t)))\\
&\leq\sum_{k=1}^{n}\frac{c}{2\varepsilon}\tilde{\delta}y^k(t)^TZZ^T\tilde{\delta}y^k(t)+\sum_{k=1}^{n}\frac{c\varepsilon}{2}(\tilde{\varphi}_k(y^k(t))\\
&\quad-\tilde{\varphi}_k(\upsilon^k(t)))^TL(\tilde{\varphi}_k(y^k(t))-\tilde{\varphi}_k(\upsilon^k(t)))\\
&=\sum_{k=1}^{n}\frac{c}{2\varepsilon}\tilde{\delta}y^k(t)^TZZ^T\tilde{\delta}y^k(t)-\frac{c\varepsilon}{2}\sum_{k=1}^{n}\sum_{i>j}l_{ij}\\
&\quad\times(\tilde{\varphi}_k(y_i^k(t))-\tilde{\varphi}_k(\upsilon_{\mu_i}^k(t))-(\tilde{\varphi}_k(y_j^k(t))-\tilde{\varphi}_k(\upsilon_{\mu_j}^k(t))))^2\\
&\leq\sum_{k=1}^{n}\frac{c}{2\varepsilon}\tilde{\delta}y^k(t)^TZZ^T\tilde{\delta}y^k(t)-c\varepsilon\sum_{k=1}^{n}\sum_{i>j}l_{ij}((\tilde{\varphi}_k(y_i^k(t))\\
&\quad-\tilde{\varphi}_k(\upsilon_{\mu_i}^k(t)))^2+(\tilde{\varphi}_k(y_j^k(t))-\tilde{\varphi}_k(\upsilon_{\mu_j}^k(t)))^2)
\end{aligned}
$$

$$
\begin{aligned}
&= \sum_{k=1}^{n} \frac{c}{2\varepsilon} \tilde{\delta y}^k(t)^T ZZ^T \tilde{\delta y}^k(t) - c\eta^2\varepsilon \sum_{k=1}^{n} \sum_{i>j} l_{ij} (\tilde{\delta y}_i^k(t)^2 + \tilde{\delta y}_j^k(t)^2) \\
&= c \sum_{k=1}^{n} \tilde{\delta y}^k(t)^T (\frac{1}{2\varepsilon} ZZ^T + \eta^2 \varepsilon (1 - \frac{1}{m}) I_m) \tilde{\delta y}^k(t). \qquad (7.10)
\end{aligned}
$$

Due to $g_k(\cdot) \in NCF(\beta, \eta)$ and $\beta > \eta$, one can get

$$
\begin{aligned}
& - \sum_{i=1}^{m} \epsilon_i \delta y_i^T(t) (G(y_i(t)) - G(\upsilon_{\mu_i}(t))) \\
&= - \sum_{k=1}^{n} \tilde{\delta y}^k(t)^T D(\tilde{g}(y^k(t)) - \tilde{g}(\upsilon^k(t))) \\
&\le -(\beta - \eta) \sum_{k=1}^{n} \tilde{\delta y}^k(t)^T D \tilde{\delta y}^k(t). \qquad (7.11)
\end{aligned}
$$

Define the matrix $M = c\beta Z + \frac{c}{2\varepsilon} ZZ^T + c\eta^2\varepsilon(1 - \frac{1}{m}) I_m$. Taking the inequalities (7.8), (7.9), (7.10) and (7.11) into account, it gives

$$
\begin{aligned}
D^+V(t) \le & \frac{1}{2} \sum_{i=1}^{m} \delta y_i(t)^T (2B_{\mu_i} + A_{\mu_i} A_{\mu_i}^T + p_{\mu_i}^2 H^T H + K_{\mu_i} K_{\mu_i}^T - 2\nu I_n) \delta y_i(t) \\
& + \frac{1}{2} \sum_{i=1}^{m} \delta y_i(t - \varrho_1(t))^T q_{\mu_i}^2 R^T R \delta y_i(t - \varrho_1(t)) \\
& + \sum_{k=1}^{n} \tilde{\delta y}^k(t)^T (M - (\beta - \eta) D + \nu I_m) \tilde{\delta y}^k(t) \\
\le & -\iota V(t) + \vartheta \sup_{\tau \in [t - \bar{\varrho}, t]} V(\tau). \qquad (7.12)
\end{aligned}
$$

Therefore, in accordance with Lemma 7.3, for $\bar{V}(t_{k-1}) = \sup_{\tau \in [t_{k-1} - \bar{\varrho}, t_{k-1}]} V(\tau)$ at the impulsive intervals $t \in [t_{k-1} - t_k)$, we have

$$
V(t) \le \bar{V}(t_{k-1}) e^{-\lambda(t - t_{k-1})}. \qquad (7.13)
$$

Considering the inequality (7.13), for the given positive parameters μ and $t > t_0 > 0$, we shall illustrate that

$$
V(t) \le \mu^{k-1} \bar{V}(t_0) e^{-\lambda(t - t_0)} \qquad (7.14)
$$

holds by introducing the mathematical induction method.

For the situation that $t \in [t_0, t_1)$ with $k = 1$, there is a positive parameter μ satisfying

$$V(t) \leq \bar{V}(t_0)e^{-\lambda(t-t_0)} = \mu^{k-1}\bar{V}(t_0)e^{-\lambda(t-t_0)}.$$

Next, assume that the inequality (7.14) holds for $t \in [t_{\tau-1}, t_\tau)$ with $k = \tau$, then we have the following result in due to condition (ii) in Theorem 7.1

$$\begin{aligned}
V(t_\tau) &\leq \Lambda V(t_\tau^-) + \Xi V(t_k^- - \varrho_2(t_\tau^-)) \\
&\leq \Lambda \mu^{\tau-1}\bar{V}(t_0)e^{-\lambda(t_\tau-t_0)} + \Xi \mu^{\tau-1}\bar{V}(t_0)e^{-\lambda(t_\tau-\varrho_2(t_\tau)-t_0)} \\
&\leq (\Lambda + \Xi e^{\lambda\varrho})\mu^{\tau-1}\bar{V}(t_0)e^{-\lambda(t_\tau-t_0)} \leq \mu^\tau \bar{V}(t_0)e^{-\lambda(t_\tau-t_0)}.
\end{aligned}$$

Finally, we will prove the inequality (7.14) holds for $k = \tau + 1$, i.e., $t \in [t_\tau, t_{\tau+1})$. From the above analysis, one can obtain

$$\begin{aligned}
V(t) &\leq \bar{V}(t_\tau)e^{-\lambda(t-t_\tau)} = \sup_{\varpi \in [t_\tau - \bar{\varrho}, t_\tau]} V(\varpi)e^{-\lambda(t-t_\tau)} \\
&= \max\{\sup_{\varpi \in [t_\tau - \bar{\varrho}, t_\tau)} V(\varpi), V(t_\tau)\}e^{-\lambda(t-t_\tau)} \\
&\leq \max\{e^{\lambda\bar{\varrho}}, \mu\}\mu^{\tau-1}\bar{V}(t_0)e^{-\lambda(t-t_0)} \leq \mu^\tau \bar{V}(t_0)e^{-\lambda(t-t_0)}.
\end{aligned}$$

Above all, we have verified the correctness of the inequality (7.14) for any $t \in [t_{k-1}, t_k)$, $k \in \mathfrak{N}^+$.

Considering that μ is a positive parameter, in the following, based on the concept of the average impulsive interval, we study the global and exponential cluster synchronization of the nonlinearly coupled Lur'e networks by classifying parameter μ.

Firstly, if $\mu \in (0, 1)$, then for $t \in [t_{k-1}, t_k)$, $k \in \mathfrak{N}^+$, it follows that

$$\begin{aligned}
V(t) &\leq \mu^{k-1}\bar{V}(t_0)e^{-\lambda(t-t_0)} \leq \mu^{\frac{t-t_0}{\hat{P}} - \bar{P}}\bar{V}(t_0)e^{-\lambda(t-t_0)} \\
&= \mu^{-\bar{P}}\bar{V}(t_0)e^{-(\lambda - \frac{\ln\mu}{\hat{P}})(t-t_0)}.
\end{aligned} \tag{7.15}$$

Secondly, if $\mu \in (1, +\infty)$, then for $t \in [t_{k-1}, t_k)$, $k \in \mathfrak{N}^+$, it implies that

$$\begin{aligned}
V(t) &\leq \mu^{k-1}\bar{V}(t_0)e^{-\lambda(t-t_0)} \leq \mu^{\frac{t-t_0}{\hat{P}} + \bar{P}}\bar{V}(t_0)e^{-\lambda(t-t_0)} \\
&= \mu^{\bar{P}}\bar{V}(t_0)e^{-(\lambda - \frac{\ln\mu}{\hat{P}})(t-t_0)}.
\end{aligned} \tag{7.16}$$

Specially, when $\mu = 1$, for $t \in [t_{k-1}, t_k)$, we have

$$V(t) \leq \mu^{k-1}\bar{V}(t_0)e^{-\lambda(t-t_0)} = \bar{V}(t_0)e^{-(\lambda - \frac{\ln\mu}{\hat{P}})(t-t_0)}. \tag{7.17}$$

As a consequence, we have three results (7.15), (7.16) and (7.17). According to Definition 7.1, it further implies that there exist constants $\lambda > 0$ and $M_0 > 0$ satisfying the following inequality

$$\|\delta y_i(t)\| \leq W_0 \bar{V}(t_0) e^{-\frac{1}{2}(\lambda - \frac{\ln \mu}{\hat{P}})(t-t_0)}.$$

Namely, the trivial solution of the error Lur'e networks (7.6) is globally and exponentially stable with the convergence rate $\frac{1}{2}(\lambda - \frac{\ln \mu}{\hat{P}})$. Moreover, the exponential cluster synchronization between the coupled Lur'e networks (7.1) and the target Lur'e system (7.2) is finally achieved under the delayed impulsive pinning controllers (7.3). We complete the proof now. □

Remark 7.2 As we can see from $\{\Lambda + \Xi e^{\lambda \varrho}, e^{\lambda \bar{\varrho}}\} \leq \mu$ in condition (ii) of Theorem 7.1, the positive parameter μ is not only directly related to the impulsive effects α, ω but also in connection to the upper bounds of system delay ϱ and impulsive delay $\bar{\varrho}$. In this regard, the smaller impulsive effects α and ω will bring a smaller μ. In some previous works like [28–30], the impulsive effects for network synchronization have been discussed into two parts: the positive one and the negative one, leading more conservative results. Additionally, the limited impulsive effect will lead to higher requirements for control performance. In this chapter, regardless of the impulsive effects α and ω are advantageous for synchronization or disadvantageous, the uniform synchronization conditions are obtained.

7.3.2 *Synchronization for Delayed Lur'e Networks*

In this subsection, we will discuss another main result of this chapter which guarantees the cluster synchronization of the nonlinearly and nonidentically coupled Lur'e networks with multiple time-varying delays.

$$\begin{aligned}\dot{y}_i(t) &= B_{\mu_i} y_i(t) + A_{\mu_i} \bar{f}^1_{\mu_i}(H y_i(t)) + K_{\mu_i} \bar{f}^2_{\mu_i}(R y_i(t - \varrho_1(t))) \\ &\quad + c \sum_{j=1}^{m} z_{ij} \Gamma G(y_j(t - \varrho_3(t))) + \bar{u}_i(t),\end{aligned} \tag{7.18}$$

where $i = 1, 2, \ldots, m$ and $\varrho_3(t)$ is the communication time-varying delay. Similarly, we design the following impulsive adaptive pinning controller

$$\bar{u}(t) = u_i^0(t) + \begin{cases} -\epsilon_i(t)(G(y_i(t)) - G(\upsilon_{\mu_i}(t))) \\ \quad -c \sum_{j=1}^{m} z_{ij} G(\upsilon_{\mu_j}(t - \varrho_3(t))), \; i \in \bar{U}_{\mu_i}, \\ 0, \quad i \in U_{\mu_i} \setminus \bar{U}_{\mu_i}, \end{cases} \tag{7.19}$$

where $\epsilon_i(t)$ is the time-varying control strength. It is assumed that $\epsilon_i(t)$ is a nonnegative monotone increasing function in this chapter.

Thus, we derive the following controlled error Lur'e networks with multiple time-varying delays

$$\begin{cases} \delta\dot{y}_i(t) = B_{\mu_i}\delta y_i(t) + A_{\mu_i}\bar{f}^1_{\mu_i}(H\delta y_i(t)) \\ \qquad +K_{\mu_i}\bar{f}^2_{\mu_i}(R\delta y_i(t-\varrho_1(t))) + c\sum_{j=1}^{m} z_{ij} \\ \qquad \times(G(y_j(t-\varrho_3(t)))-G(\upsilon_{\mu_j}(t-\varrho_3(t)))) \\ \qquad -\epsilon_i(t)(G(y_i(t)) - G(\upsilon_{\mu_i}(t))), \quad t \neq t_k, \\ \Delta\delta y_i(t_k) = \alpha\delta y_i(t_k^-) + \omega\delta y_i(t-\varrho_2(t_k^-)), \quad t = t_k, \end{cases} \tag{7.20}$$

where $\delta y_i(t)$ is presumed to be right-hand continuous at $t = t_k$, $k \in \mathfrak{N}^+$, and $\delta y_i(t_k) = \delta y_i(t_k^+) = \lim_{t\to t_k^+}\delta y_i(t)$, $\delta y_i(t_k^-) = \lim_{t\to t_k^-}\delta y_i(t)$. Similar to the previous section, the following theorem is raised to ensure the realization of the cluster synchronization.

Theorem 7.2 *Suppose that Assumption 7.1 holds, the average impulsive interval is no larger than $\hat{P}$ for the impulsive sequence $\zeta = \{t_1, t_2, \ldots\}$ and $g_k(\cdot) \in NCF(\beta, \eta)(k = 1, 2, \ldots, n)$ for positive constants $\beta > \eta$. For the positive parameter h_i, design the adaptive updating laws as*

$$\dot{\epsilon}_i(t) = h_i\delta y_i^T(t)(G(y_i(t)) - G(\upsilon_{\mu_i}(t))), \quad i = 1, 2, \ldots, m. \tag{7.21}$$

If there exist positive parameters c, $\acute{\eta}$, ν, ε and the control strength matrix D, such that
(i) *The matrix inequality*

$$\Omega = \begin{bmatrix} \eta I_m + \frac{c}{2\varepsilon}ZZ^T - (\beta-\eta)D & \frac{1}{2}c\beta Z \\ * & -\acute{\eta}I_m \end{bmatrix} \leq 0, \tag{7.22}$$

(ii) *For positive constant $\acute{\mu}$, there holds*

$$\{\Lambda + \Xi e^{\acute{\lambda}\varrho}, e^{\acute{\lambda}\tilde{\varrho}}\} \leq \acute{\mu}, \tag{7.23}$$

then the controlled error Lur'e networks (7.20) are globally and exponentially stable with the convergence rate $\frac{1}{2}(\acute{\lambda} - \frac{\ln\acute{\mu}}{\hat{P}})$, where $\Lambda = (1+\alpha)(1+\alpha+\omega)$, $\Xi = \omega(1+\alpha+\omega)$ and $\acute{\lambda}$ is the unique feasible solution to the equation $\acute{\lambda} - \iota + \bar{\vartheta}e^{\acute{\lambda}\tilde{\varrho}} = 0$ with parameters $\iota = -\min_{1\leq i\leq m}\{\lambda_{max}(B^s_{\mu_i} + A_{\mu_i}A^T_{\mu_i} + p^2_{\mu_i}H^TH + K_{\mu_i}K^T_{\mu_i} - 2\nu I_n)\}$, $\vartheta = \max_{1\leq i\leq m}\{\lambda_{max}(q^2_{\mu_i}R^TR)\}$, $\acute{\vartheta} = \max_{1\leq i\leq m}\{\lambda_{max}(2c\eta^2\varepsilon(1-\frac{1}{m})I_n + 2\acute{\eta}I_n)\}$, $\tilde{\varrho} = \max\{\varrho_1(t), \varrho_3(t)\}$ satisfying $\bar{\vartheta} = \vartheta + \acute{\vartheta}$ and $\iota \leq \vartheta$. That is, the global and exponential cluster synchronization of the nonlinearly coupled Lur'e networks (7.18) is finally obtained by importing the impulsive pinning controllers (7.19) and the adaptive updating laws (7.21).

Proof Construct the following Lyapunov function

$$V(t) = \frac{1}{2}\sum_{i=1}^{m} \delta y_i(t)^T \delta y_i(t) + \sum_{i=1}^{m} \frac{1}{2h_i}(\epsilon_i(t) - \epsilon_i)^2.$$

For the impulsive instants $t = t_k$, $k \in \mathfrak{N}^+$, from the second equation of (7.20), one may obtain

$$\begin{aligned} V(t_k^+) &= \frac{1}{2}\sum_{i=1}^{m} \delta y_i(t_k^+)^T \delta y_i(t_k^+) + \sum_{i=1}^{m} \frac{1}{2h_i}(\epsilon_i(t_k^+) - \epsilon_i)^2 \\ &\leq \frac{1}{2}(1+\alpha)(1+\alpha+\omega)\delta y(t_k^-)^T \delta y(t_k^-) \\ &\quad + \sum_{i=1}^{m} \frac{1}{2h_i}(\epsilon_i(t_k^-) - \epsilon_i)^2 + \frac{1}{2}\omega(1+\alpha+\omega) \\ &\quad \times \delta y(t_k^- - \varrho_2(t_k^-)))^T \delta y(t_k^- - \varrho_2(t_k^-)) \\ &\quad + \sum_{i=1}^{m} \frac{1}{2h_i}(\epsilon_i(t_k^- - \varrho_2(t_k^-)) - \epsilon_i)^2 \\ &= \Lambda V(t_k^-) + \Xi V(t_k^- - \varrho_2(t_k^-)). \end{aligned} \tag{7.24}$$

Calculating the derivative of $V(t)$ along the trajectory of the controlled error Lur'e networks (7.20) for $t \in (t_{k-1}, t_k)$, $k \in \mathfrak{N}^+$ gives

$$\begin{aligned} D^+V(t) &\leq \frac{1}{2}\sum_{i=1}^{m} \delta y_i(t)^T (2B_{\mu_i} + A_{\mu_i}A_{\mu_i}^T + p_{\mu_i}^2 H^T H + K_{\mu_i}K_{\mu_i}^T)\delta y_i(t) \\ &\quad + \frac{1}{2}\sum_{i=1}^{m} \delta y_i(t-\varrho_1(t))^T q_{\mu_i}^2 R^T R \delta y_i(t-\varrho_1(t))) \\ &\quad + c\sum_{i=1}^{m}\sum_{j=1}^{m} z_{ij}\delta y_i(t)^T (G(y_j(t-\varrho_3(t))) - G(\upsilon_{\mu_j}(t-\varrho_3(t)))) \\ &\quad - \sum_{i=1}^{m} \epsilon_i(t)\delta y_i(t)^T (G(y_i(t)) - G(\upsilon_{\mu_i}(t)))) \\ &\quad + \sum_{i=1}^{m} (\epsilon_i(t) - \epsilon_i)\delta y_i(t)^T (G(y_i(t)) - G(\upsilon_{\mu_i}(t)))). \end{aligned} \tag{7.25}$$

Similarly, it shows that

$$c\sum_{i=1}^{m}\sum_{j=1}^{m} z_{ij}\delta y_i(t)^T(G(y_j(t-\varrho_3(t)))-G(\upsilon_{\mu_j}(t-\varrho_3(t))))$$
$$\leq c\beta\sum_{k=1}^{n}\tilde{\delta} y^k(t)^T Z\tilde{\delta} y^k(t-\varrho_3(t))+\sum_{k=1}^{n}\frac{c}{2\varepsilon}\tilde{\delta} y^k(t)^T ZZ^T\tilde{\delta} y^k(t)$$
$$+\ c\eta^2\varepsilon\sum_{i=1}^{m}(1-\frac{1}{m})\delta y_i(t-\varrho_3(t))^T\delta y_i(t-\varrho_3(t)),$$

where $\tilde{\delta} y^k(t-\varrho_3(t))=[\delta y_1^k(t-\varrho_3(t)),\delta y_2^k(t-\varrho_3(t)),\ldots,\delta y_m^k(t-\varrho_3(t))]^T$. Then, let $\xi=(\tilde{\delta} y^k(t)^T,\tilde{\delta} y^k(t-\varrho_3(t))^T)$, we have

$$
\begin{aligned}
D^+V(t) &\leq \frac{1}{2}\sum_{i=1}^{m}\delta y_i(t)^T(2B_{\mu_i}+A_{\mu_i}A_{\mu_i}^T\\
&\quad+\ p_{\mu_i}^2H^TH+K_{\mu_i}K_{\mu_i}^T-2\nu I_n)\delta y_i(t)\\
&\quad+\ \frac{1}{2}\sum_{i=1}^{m}\delta y_i(t-\varrho_1(t))^T q_{\mu_i}^2R^TR\delta y_i(t-\varrho_1(t))\\
&\quad+\ \frac{1}{2}\sum_{i=1}^{m}\delta y_i(t-\varrho_3(t))^T(2c\eta^2\varepsilon(1-\frac{1}{m})I_n\\
&\quad+\ 2\acute{\eta}I_n)\delta y_i(t-\varrho_3(t))+\sum_{k=1}^{n}\xi\Omega\xi^T\\
&\leq -\iota(\sum_{k=1}^{n}\tilde{\delta} y^k(t)^T\tilde{\delta} y^k(t)+\sum_{i=1}^{m}\frac{1}{2h_i}(\epsilon_i(t)-\epsilon_i)^2)\\
&\quad+\ \vartheta(\frac{1}{2}\sum_{i=1}^{m}\delta y_i(t-\varrho_1(t))^T\delta y_i(t-\varrho_1(t))\\
&\quad+\ \sum_{i=1}^{m}\frac{1}{2h_i}(\epsilon_i(t)-\epsilon_i)^2)\\
&\quad+\ \acute{\vartheta}(\frac{1}{2}\sum_{i=1}^{m}\delta y_i(t-\varrho_3(t))^T\delta y_i(t-\varrho_3(t))\\
&\quad+\ \sum_{i=1}^{m}\frac{1}{2h_i}(\epsilon_i(t)-\epsilon_i)^2)\\
&\leq -\iota V(t)+\vartheta V(t-\varrho_1(t))+\acute{\vartheta}V(t-\varrho_3(t))\\
&\leq -\iota V(t)+\bar{\vartheta}\sup_{\tau\in[t-\tilde{\varrho},t]}V(\tau),
\end{aligned}
\tag{7.26}
$$

where $\iota = -\min_{1\le i\le m}\{\acute{\lambda}_{\max}(B^s_{\mu_i} + A_{\mu_i}A^T_{\mu_i} + p^2_{\mu_i}H^TH + K_{\mu_i}K^T_{\mu_i} - 2\nu I_n)\}$, $\vartheta = \max_{1\le i\le m}\{\acute{\lambda}_{\max}(q^2_{\mu_i}R^TR)\}$, $\acute{\vartheta} = \max_{1\le i\le m}\{\acute{\lambda}_{\max}(2c\eta^2\varepsilon(1-\frac{1}{m})I_n + 2\acute{\eta}I_n)\}$, $\bar{\vartheta} = \vartheta + \acute{\vartheta}$, $\iota \le \bar{\vartheta}$, $\tilde{\varrho} = \max\{\varrho_1(t), \varrho_3(t)\}$ and $\sum_{i=1}^m \frac{1}{h_i}(\epsilon_i(t) - \epsilon_i)\dot{\epsilon}_i(t) \le 0$. Similarly, the cluster synchronization of the nonlinearly coupled Lur'e networks is realized under the impulsive pinning controller (7.19). Based on adaptive updating laws (7.21), some suitable feedback control strengths are obtained. So far, the proof of Theorem 7.2 is completed. □

Remark 7.3 In this chapter, to achieve the cluster synchronization of the coupled Lur'e networks, it is assumed that $\epsilon_i(t)$ is a nonnegative monotone increasing function. Based on Lyapunov stability theorem, it is obvious that we will finally have $V(t) \to 0 (t \to \infty)$. From the perspective of the selection on Lyapunov function, there exists $\epsilon_i(t) \to \epsilon_i (i = 1, 2, \ldots, m)$. That is, the derivative of the function $\sum_{i=1}^m \frac{1}{2h_i}(\epsilon_i(t) - \epsilon_i)^2$ is $\sum_{i=1}^m \frac{1}{h_i}(\epsilon_i(t) - \epsilon_i)\dot{\epsilon}_i(t) < 0$, which guarantees the validation of the result (7.26) for each continuous interval $t \in [t_{k-1}, t_k)$ with $k \in \mathfrak{N}^+$. Different from most previous results [22, 31, 32], the adaptive control method was applied in this chapter, which is a better fit for the practical conditions. In case that if the above assumption was not rendered, the compact form $D^+ \le \iota V(t) + \bar{\vartheta}\sup_{\tau\in[t-\tilde{\varrho},t]} V(\tau)$ could not be derived. Additionally, the mathematical induction could not be applied directly in this proof. Above all, the assumption on $\epsilon_i(t)$ is essential.

Remark 7.4 In the previous papers like [19, 29, 33–35], the impulsive intervals were roughly set as $\inf_{k\in\mathfrak{N}^+}\{t_k - t_{k-1}\}$ or $\sup_{k\in\mathfrak{N}^+}\{t_k - t_{k-1}\}$, which finally needs a tight choice of the controller parameters and results in wasting of resources. In fact, based on the concept of the average impulsive interval in Definition 7.2 in this chapter, the impulsive times of the impulsive sequence ζ with impulsive interval (t, s) could be estimated by adjusting two positive parameters $\hat{P}$ and $\bar{P}$. For the parameter $\hat{P}$ which is related to impulsive intervals, we have $\inf_{k\in\mathfrak{N}^+}\{t_k - t_{k-1}\} \le \hat{P} \le \sup_{k\in\mathfrak{N}^+}\{t_k - t_{k-1}\}$. On this point, by employing the definition of the average impulsive interval, some less conservative results can be gained.

Remark 7.5 Due to the variability of the actual environments, time delay has become one of the most important uncertain factors that must be considered in modeling the networks. In order to conform in practical situations, the case of the system delay $\varrho_1(t)$, the impulsive state delay $\varrho_2(t)$ and the coupling delay $\varrho_3(t)$ are discussed in this chapter, which is more thoughtful than the existing papers [3, 36]. According to the conditions (7.22) and (7.23) in the Theorem 7.2, it can be found that the synchronization criteria are delay-dependent. That is, the upper bounds of the system time-varying delay, the impulsive state delay and coupling delay are required to be known. Additionally, from the definition of exponential convergence rate in Theorem 7.2, it obtains that the convergence rate $\acute{\lambda}$ is relevant to the upper bounds of three time-varying delays. Simultaneously, according to condition (7.23), there is an obvious relation that the smaller the upper bounds of three time-varying delays will bring the faster convergence rate.

7.4 Numerical Simulations

We construct the Chua's circuits with different system parameters [37]

$$\dot{y}(t) = B_i y(t) + A_i f_i^1(H_i y(t)) + K_i f_i^2(R_i y(t - \varrho_i(t))),$$

where

$$y(t) = \begin{bmatrix} y^1(t) \\ y^2(t) \\ y^3(t) \end{bmatrix}, \quad A_1 = A_2 = \begin{bmatrix} 3.247 \\ 0 \\ 0 \end{bmatrix},$$

$$K_1 = K_2 = \begin{bmatrix} 0 \\ 0 \\ -3.906 \end{bmatrix},$$

$$B_1 = \begin{bmatrix} -\frac{13}{6} & 10 & 0 \\ 1 & -1 & 1 \\ 0 & -19.53 & -0.201 \end{bmatrix},$$

$$B_2 = \begin{bmatrix} -2.169 & 10 & 0 \\ 1 & -1 & 1 \\ 0 & -19.53 & -0.1636 \end{bmatrix},$$

$H_i = R_i = [1, 0, 0]$, $f_i^1(H_i y(t)) = (|y^1(t) + 1| - |y^1(t) - 1|)$ and $f_i^2(R_i y(t)) = \sin(0.5 y^1(t - \varrho_i(t)))$, where $\varrho_1 = \frac{0.01e^t}{1+e^t}$ and $\varrho_2 = \frac{0.2e^t}{1+e^t}$. The phase portrait of the two Chua's circuits are shown in Fig. 7.1a.

Example 7.1 In this numerical example, consider the nonlinearly coupled networks consisted of six Lur'e systems with two clusters $U_1 = \{1, 2, 3\}$ and $U_2 = \{4, 5, 6\}$. According to Fig. 7.1b and pinning control scheme, the second and the fourth Lur'e systems are selected to be controlled. Let $c = 0.5$, $Z = [-1, 1, 0, 0, 0, 0; 0, -2, 1, 1, 0, 0; 0, 1, -1, 0, 0, 0; 0, 0, 0, -1, 1, 0; 0, 0, 0, 0, -1, 1; 0, 0, 0, 0, 1, -1]$, and $G(y_i(t - \varrho_3(t))) = y_i(t - \varrho_3(t)) + 0.2\sin(y_i(t - \varrho_3(t)))$ and the time-varying delay $\varrho_3(t) = 0.3 + 0.1\sin(0.2t)$. In addition, the impulsive sequence $\zeta = \{t_1, t_2, \ldots\}$ is given in Fig. 7.2a, of which we suppose the average impulsive interval $\hat{P}$ is less than 0.02.

Define the synchronization error of the two clusters as

$$E_j(t) = \sqrt{\frac{1}{3}\sum_{i=1}^{3} \|y_i(t) - \upsilon_{\mu_i}(t)\|}, \ j = 1, 2.$$

Then we have the evolution curves of error in Fig. 7.2b. In accordance with Definition 7.1, it is obvious the cluster synchronization is realized. Define the state error between two clusters as $E_{12}(t) = \sqrt{1/3((y_1(t))^2 + (y_2(t))^2 + (y_3(t))^2)} - \sqrt{1/3((y_4(t))^2 + (y_5(t))^2 + (y_6(t))^2)}$. It can be found in Figs. 7.2b and 7.3a that the

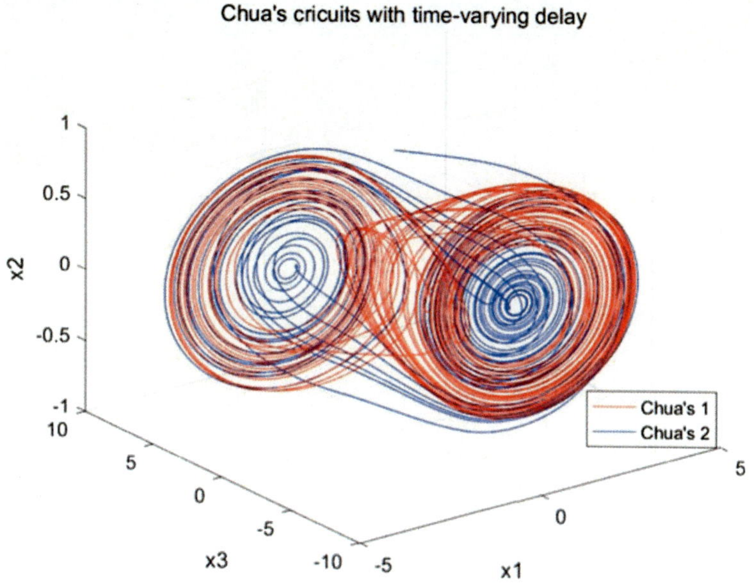

(a)

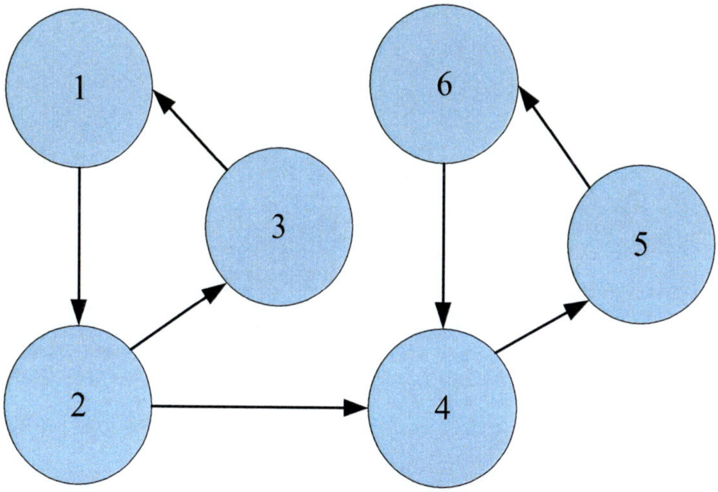

(b)

Fig. 7.1 **a** The phase portrait of two Chua's circuits. **b** The topological structure of the Lur'e networks

error curves in the same cluster approach to zero as time goes to 0.15s, but the error $E_{12}(t)$ show irregular oscillations. Referring to Fig. 7.3b, the control strength matrix $D = \text{diag}\{0, 0.5155, 0, 0.44, 0, 0\}$ is easy to get. Through the above pictures, it can be concluded that this simulation shows the effectiveness of the theorem well.

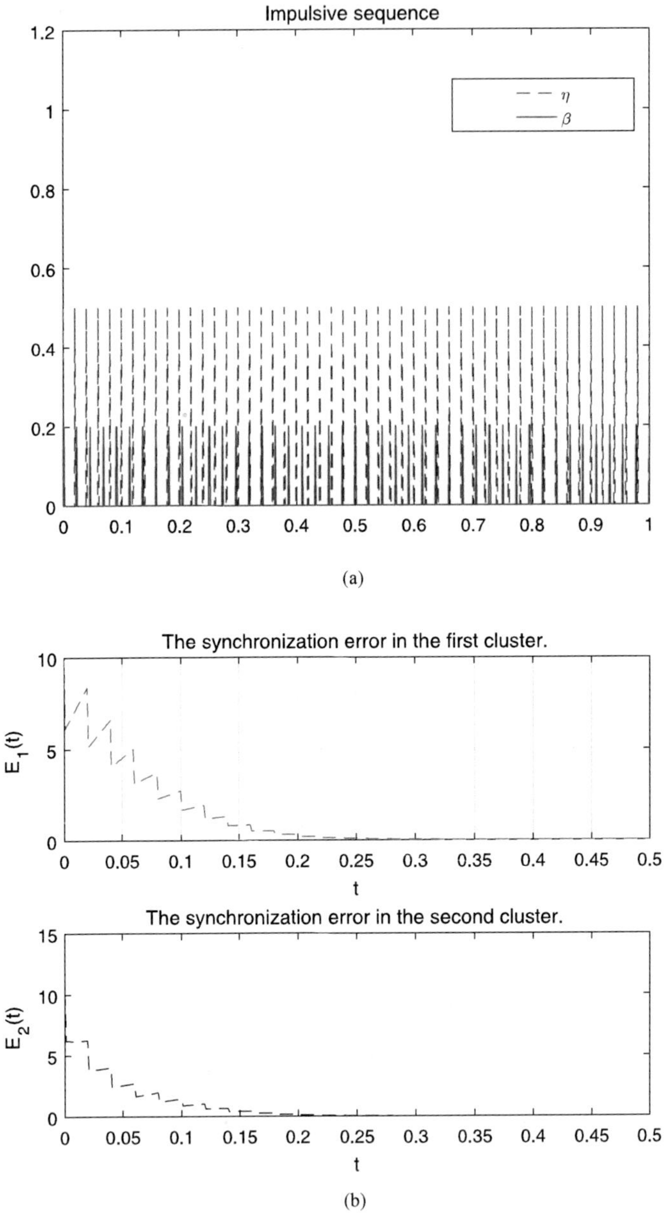

Fig. 7.2 **a** The impulsive sequence $\zeta = \{t_1, t_2, \ldots\}$. **b** Synchronization error $E_j(t)(j = 1, 2)$ in each cluster

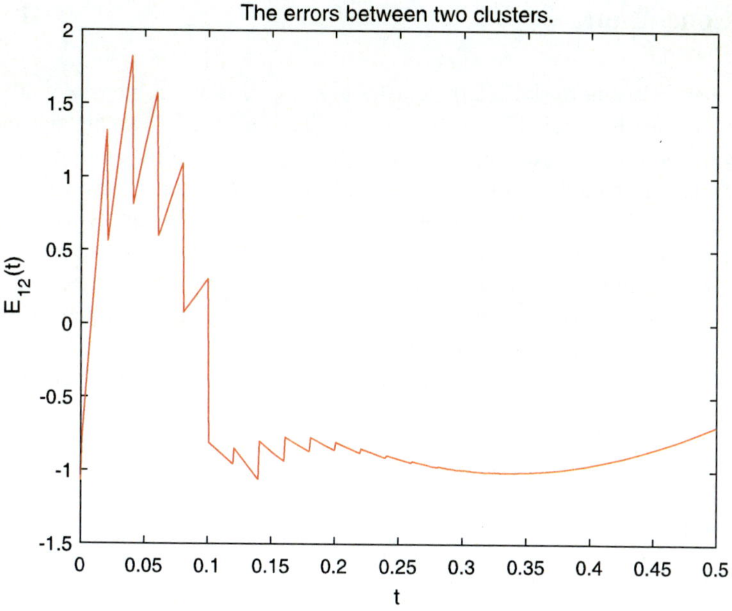

(a)

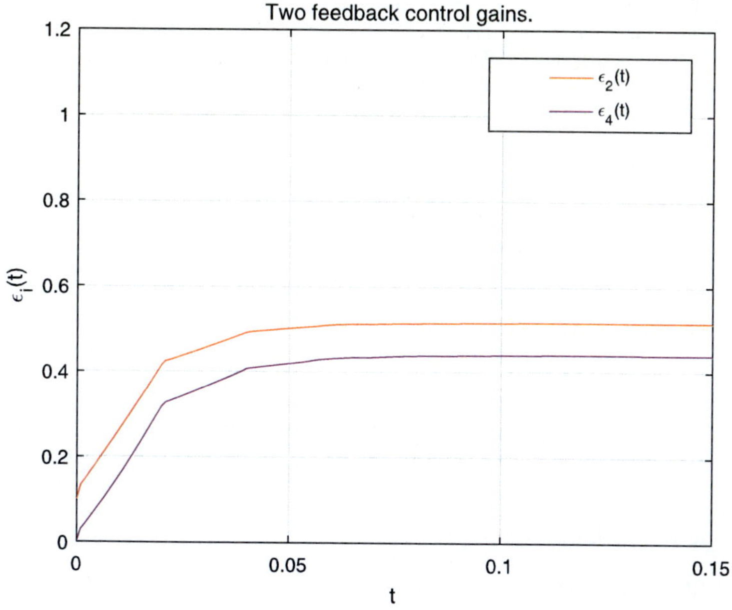

(b)

Fig. 7.3 **a** The error curves between two clusters. **b** The curves of two control gains strengths

7.5 Conclusion

In this chapter, we have studied the cluster synchronization problem of nonlinearly and non identically coupled Lur'e networks with multiple time-varying delays. With the introduction of average impulsive interval, Lyapunov stability theorem and mathematical induction method, we have obtained sufficient conditions for achieving the exponentially cluster synchronization of the Lur'e networks by applying the impulsive pinning controller. In addition, on account of the designing adaptive updating laws, effective and comparatively small control strengths are obtained, which are more economical than the given control strengths. Finally, a given numerical simulation has verified the effectiveness of the theoretical results and the proposed control schemes. In the following, we will try to investigate impulsive Lur'e networks with multiple time-varying delays and exogenous disturbances.

References

1. Strogatz, S.H.: Exploring complex networks. Nature. **410**(6825), 268–276 (2001)
2. He, W.L., Qing, F., Cao, J.D.: Pinning-controlled synchronization of delayed neural networks with distributed-delay coupling via impulsive control. Neural Networks. **85**, 1–9 (2017)
3. Ding, D., Tang, Z., Wang, Y.: Synchronization of nonlinearly coupled complex networks: distributed impulsive method. Chaos, Solitons and Fractals. **133**, 109620 (2020)
4. Yu, T.H., Cao, D.Q., Yang, Y.: Robust synchronization of impulsively coupled complex dynamical network with delayed nonidentical nodes. Chaos, Solitons and Fractals. **87**, 92–101 (2016)
5. Wang, Y.Q., Lu, J.Q., Ling, J.L.: Pinning synchronization of nonlinear coupled Lur'e networks under hybrid impulses. IEEE Transactions on Circuits and Systems II: Express Briefs. **66**(3), 432–436 (2019)
6. Chen, T.P., Liu, X., Lu, W.: Pinning complex networks by a single controller. IEEE Transactions on Circuits and Systems I: Regular Papers. **54**(6), 1317–1326 (2007)
7. Yadav, V.K., Shukla, V.K., Das, S.: Difference synchronization among three chaotic systems with exponential term and its chaos control. Chaos, Solitons and Fractals. **124**, 36–51 (2019)
8. Estrada, E., Ganbuzza, L.V., Frasca, M.: Synchronization in networks with heterogeneous coupling delays. SIAM Journal on Applied Dynamical Systems. **17**(1), 672–693 (2018)
9. Yu, W.W., Cao, J.D., Chen, G.R.: Local synchronization of a complex network model. IEEE Transactions on Systems, Man, and Cybernetics, Part B. **39**(1), 230–241 (2009)
10. Li, C.G., Chen, G.R.: Synchronization in general complex dynamical networks with coupling delays. Physical A: Statistical Mechanics and its Applications. **343**, 263–278 (2004)
11. Gao, L.K., Xiao, J., Guo, X.X.: Adaptive variable impulsive synchronization of multi-agent systems. Journal of Electrical Engineering and Technology. **15**(1), 259–268 (2020)
12. Feng, J.W., Tang, Z., Zhao, Y.: Cluster synchronisation of non-linearly coupled Lur'e networks with identical and non-identical nodes and an asymmetrical coupling matrix. IET Control Theory and Applications. **7**(18), 2117–2127 (2013)
13. Ma, Z.J., Liu, Z.R., Zhang, G.: A new method to realize cluster synchronization in connected chaotic networks. Chaos: An Interdisciplinary Journal of Nonlinear Science. **16**(2), 023103 (2006)
14. Wu, W., Zhou, W.J., Chen, T.P.: Cluster synchronization of linearly coupled complex networks under pinning control. IEEE Transactions on Circuits and Systems I: Regular Papers. **56**(4), 829–839 (2009)

15. Li, X.F., Chu, Y.D., Leung, A.Y.T.: Synchronization of uncertain chaotic systems via complete-adaptive-impulsive controls. Chaos, Solitons and Fractals. **100**, 24–30 (2017)
16. Lu, J.Q., Kurths, J., Cao, J.D.: Synchronization control for nonlinear stochastic dynamical networks: pinning impulsive strategy. IEEE Transactions on Neural Networks and Learning Systems. **23**(2), 285–292 (2012)
17. Xu, C., Tong, D.B., Chen, Q.Y.: Exponential synchronization of chaotic systems with stochastic noise via periodically intermittent control. International Journal of Robust and Nonlinear Control. **30**(7), 2611–2624 (2020)
18. Tang, Z., Park, J.H., Wang, Y.: Impulsive synchronization of derivative coupled neural networks with cluster-tree topology. IEEE Transactions on Network Science and Engineering. **7**(3), 1788–1798 (2020)
19. Chen, H.B., Shi, P., Lim, C.C.: Pinning impulsive synchronization for stochastic reaction-diffusion dynamical networks with delay. Neural Networks. **106**, 281–293 (2018)
20. Wang, J.Y., Feng, J.W., Xu, C.: Cluster synchronization of nonlinearly-coupled complex networks with nonidentical nodes and asymmetrical coupling matrix. Nonlinear Dynamics. **67**(2), 1635–1646 (2012)
21. Lu, J.Q., Ho, D.W.C., Cao, J.D.: A unified synchronization criterion for impulsive dynamical networks. Automatica. **46**(7), 1215–1221 (2010)
22. Kumar, R., Sarkar, S., Das, S.: Projective synchronization of delayed neural networks with mismatched parameters and impulsive effects. IEEE Transactions on Neural Networks and Learning Systems. **31**(4), 1211–1221 (2020)
23. Wu, W., Zhou, W.J., Chen, T.P.: Cluster synchronization of linearly coupled complex networks under pinning control. IEEE Transactions on Circuits and Systems. Part I: Regular Papers. **56**(4), 829–839 (2009)
24. He, X.Y., Peng, D.X., Li, X.D.: Synchronization of complex networks with impulsive control involving stabilizing delay. Journal of the Franklin Institute. **357**(8), 4869–4886 (2020)
25. Liu, X., Chen, T.P.: Synchronization analysis for nonlinearly-coupled complex networks with an asymmetrical coupling matrix. Physical A: Statistical Mechanics its Applications. **387**, 4429–4439 (2008)
26. Tang, Z., Feng, J.W., Zhao, Y.: Global synchronization of nonlinear coupled complex dynamical networks with information exchanges at discrete-time. Neurocomputing. **151**(mar.3pt.3), 1486–1494 (2015)
27. Bebernes, J.: Differential equations stability, oscillations, time lags. SIAM Review. **10**(1), 93–94 (1968)
28. Xu, W.Y., Ho, D.W.C.: Clustered event-triggered consensus analysis: an impulsive framework. IEEE Transactions on Industrial Electronics. **63**(11), 7133–7143 (2016)
29. Pu, H., Liu, Y., Jang, H.: Exponential synchronization for fuzzy cellular neural networks with time-varying delays and nonlinear impulsive effects. Cognitive Neurodynamics. **9**(4), 437–446 (2015)
30. Li, X.D., Song, S.J.: Impulsive control for existence, uniqueness, and global stability of periodic solutions of recurrent neural networks with discrete and continuously distributed delays. IEEE transactions on neural networks and learning systems. **24**(6), 868–877 (2013)
31. Tang, Z., Park, J.H., Zheng, W.X.: Distributed impulsive synchronization of Lur'e dynamical networks via parameter variation methods. International Journal of Robust and Nonlinear Control. **28**(3), 1001–1015 (2018)
32. Yan, H.L., Zhong, S.M., Shu, L.: Synchronization of nonlinear complex dynamical systems via delayed impulsive distributed control. Applied Mathematics and Computation. **320**, 75–85 (2018)
33. Yan, H.L., Liu, Z.W., Feng, G.: Synchronization of complex dynamical networks with time-varying delays via impulsive distributed control. IEEE Transactions on Circuits and Systems I: Regular Papers. **157**(8), 2182–2195 (2010)
34. Zhang, W., Huang, J.J., Wei, P.C.: Weak synchronization of chaotic neural networks with parameter mismatch via periodically intermittent control. Applied Mathematical Modelling. **35**(2), 612–620 (2011)

35. Hu, J.Q., Liang, J.L., Cao, J.D.: Synchronization of hybrid-coupled heterogeneous networks: pinning control and impulsive control schemes. Journal of the Franklin Institute. **351**(5), 2600–2622 (2014)
36. Wang, X., Liu, X.Z., She, K.: Delay-dependent impulsive distributed synchronization of stochastic complex dynamical networks with time-varying delays. IEEE Transactions on Systems, Man, and Cybernetics: Systems. **49**(7), 1496–1504 (2019)
37. Yalcin, M.E., Suykens, J.A.K, Vandewalle, J.: Master-slave synchronization of Lur'e systems with time delay. International Journal of Bifurcation and Chaos in Applied Sciences and Engineering. **11**(6), 1707–1722 (2001)

Chapter 8
Synchronization of Derivative Coupled CDNs with Hybrid Impulses

8.1 Introduction

Recent years have witnessed the rapid development of complex networks, including link prediction [1], network structure [2], synchronization dynamics [3], etc., which play an significant role in transportation [4], energy networks [5], and information security [6]. Among many dynamic behaviors in complex networks, synchronization has attracted the attention from a large number of scholars due to its unique perspective on the collective behavior of the nodes in a complex network. Synchronization is the study of realizing states of nodes with interconnected and similar dynamic properties to be consistent/almost consistent. Generally speaking, there are two ways to achieve synchronization, that is, depending on the mutual influence of couplings between complex networks to achieve synchronization, or imposing control signals to the nodes to achieve synchronization. In [7], it is proposed for the first time that synchronization of coupled complex networks was realized through the mutual influence of linear coupling. After that, synchronization of a kind of nonlinearly coupled complex networks was investigated in [8]. Through the linearization process of NCF, it turned out that synchronization was achieved without external control inputs. However, with the continuously deepening of researches on complex networks, more and more situations with realistic physical significance are considered such as time delays [9], nonlinearities [10] and asymmetric couplings [11]. The couplings between complex networks can sometimes be regarded as unfavorable factors, which may cause global instability. Therefore, the introducing of control inputs to help achieving synchronization is a reliable solution.

In recent years, control protocols including pinning negative feedback controller [12], intermittent controller [13], and impulsive controller [14] have been adopted to synchronize various complex networks. For example, Yu et al. designed a kind of distributed pinning controllers and successfully achieved cluster synchroniza-

Z. Tang et al., *Impulsive Synchronization of Complex Dynamical Networks*,
https://doi.org/10.1007/978-981-16-5383-4_8

tion issue of complex networks under fixed topology and switched topology [15]. Amhed et al. investigated synchronization of a complex network with multiple subnetworks by employing adaptive feedback controllers [16]. In both researches, the global synchronization were obtained by imposing control inputs. On this basis, scholars further explore methods that require less control costs. Cheng et al. considered the drive-response synchronization issue of the complex network with coupled time-varying delay and distributed delay via a periodically intermittent controller [17]. To proceed, Zhou et al. expanded the control period of the controller from a constant to a variable, i.e., aperiodically control intervals, which expanding the usage scenarios of intermittent controllers [18]. However, whether it was a negative pinning controller or an intermittent controller, both of which released continuous control signals, which was not conducive to be applied in some discrete systems. Besides, within some harsh working environments such as cyber attacks [19] or Dos attacks [20], it is quite difficult to establish stable and continuous communication channels. Therefore, impulsive control protocol which is a kind of discrete control method is preferred.

Recalling the researches in the past few years, the impulsive controller scheme has already shown its high efficiency. In [21], Peng et al. studied the leader-following synchronization with the event-triggered impulsive controller. The result illustrated that the complex network can be synchronized with only a small amount of impulse signals. Li et al. established an impulsive controller to synchronize a class of fractional-order complex networks [22]. With the expansion of Barbalet's lemma, criterion for discrete model was also given. In spite of the above results have successfully achieved synchronization through the impulsive controller, they all assumed that the impulsive effects were beneficial to the synchronization results. In fact, the impulses released by controllers could be regarded as synchronizing impulses and desynchronizing impulses according to the different ranges of impulsive effects. Obviously, analyzing the conditions in term of these two different cases will undoubtedly make the obtained results more universal. Through the discussion of the different ranges of impulsive effects, Li et al. realized the synchronization for the stochastic discrete complex networks, however, authors failed to give the specific convergence rates with respect to synchronizing and desynchronizing impulses [23]. To tackle the difficulty, parameter variation method is introduced to solve such problem. In the related works of Zhou et al. and He et al., by establishing the comparison functions and with the help of parameter variation method, the convergence rates of the complex networks were accurately given. The fly in the ointment is that [24, 25] did not consider the network models with derivative coupling.

Motivated by all the stuffs discussed above, this chapter is devoted to discussing a leader-following synchronization for the complex network with derivative coupling and multi-delay couplings. To deal with the derivative coupling involved, a novel comparison system is established. We will show how the parameter variation method is extended in this work. The contributions of this chapter mainly lies in the following three aspects:

1. To study more complicated forms of network coupling, a complex network model with derivative coupling is considered, thereby, a situation closer to reality is portrayed rather than [26, 27]. In addition, a novel Lyapunov function is proposed to deal with the derivative coupling.
2. Different with those researches only focusing on impulses with positive impulsive effects [28], the influence of synchronizing and desynchronizing impulses on synchronization are analyzed separately by taking different ranges of impulsive effects into account.
3. Based on comparison principle, contradiction analysis method, and the definition of average impulsive interval, the specific convergence rates regarding with synchronizing and desynchronizing impulses are derived. Compared with some literatures [29, 30], the results given in this chapter are less conservative.

This chapter is organized as follows. In Sect. 8.2, the model of Lur'e networks are given and some useful definitions, lemmas and assumptions are introduced. The synchronization conditions with synchronizing impulses and desynchronizing impulses are derived in Sect. 8.3. In Sect. 8.4, three numerical simulations are given. Finally, the conclusion is drawn in Sect. 8.5.

Notations. Notation $\lambda_{\max}(\cdot)$ denotes the largest eigenvalue of a matrix. The symmetrical part of $\mathcal{L}$ is represented by $\mathcal{L}^s = \frac{1}{2}(\mathcal{L} + \mathcal{L}^T)$. If matrix $\mathcal{L}$ is positive definite (semi-definite), then we denote it as $\mathcal{L} > 0 (\mathcal{L} \geq 0)$. I_n stands for the identity matrix with n dimension. For a coupling matrix $\Delta = (\delta_{ij})_{N\times N}$, if there exists a link from the i-th node to the j-th node, then $\delta_{ij} = 1$, otherwise, $\delta_{ij} = -1$. $\otimes$ is the sign for Kronecker product. $\mathcal{N}$ denotes the set of positive integers. The dimension of vectors and matrices will be stated in the context.

8.2 Network Model and Preliminaries

In this section, the model of Lur'e networks will be given. Additionally, some definitions and lemmas will be introduced.

8.2.1 Network Model

Consider the chaotic behavior of the leader system as

$$\dot{z}_\pi(t) = \Omega_{A_i} z_\pi(t) + \Omega_{B_i}\tilde{\chi}_1(\Omega_{C_i} z_\pi(t)) + \Omega_{D_i}\tilde{\chi}_2(\Omega_{E_i} z_\pi(t - \tau(t))), \tag{8.1}$$

where $z_\pi(t) = [z_\pi^1(t), z_\pi^2(t), \ldots, z_\pi^n(t)]^T \in \mathbb{R}^n$ is the state vector of the leader system; $\Omega_{A_i} \in \mathbb{R}^{n\times n}$, $\Omega_{B_i}, \Omega_{D_i} \in \mathbb{R}^{n\times m}$, $\Omega_{C_i}, \Omega_{E_i} \in \mathbb{R}^{m\times n}$ denote the positive matrices; $\tilde{\chi}_1, \tilde{\chi}_2 : \mathbb{R}^n \to \mathbb{R}^n$ represent two nonlinear functions; Time-varying delay $\tau(t)$ satisfies $0 < \tau(t) < \tau$ and $0 < \dot{\tau}(t) < 1$.

The model of Lur'e network can be expressed as

$$\begin{aligned}\dot{z}_i(t) &= \Omega_{A_i} z_i(t) + \Omega_{B_i}\tilde{\chi}_1(\Omega_{C_i} z_i(t)) + \Omega_{D_i}\tilde{\chi}_2(\Omega_{E_i} z_i(t-\tau(t))) \\ &\quad + v_1\sum_{j=1}^{N}\delta_{ij}\Gamma z_j(t) + v_2\sum_{j=1}^{N} s_{ij}\Gamma z_j(t-\omega(t)) \\ &\quad + v_3\sum_{j=1}^{N} r_{ij}\Gamma \dot{z}_j(t) + u_i(t), \qquad i = 1, 2, \ldots, N, \end{aligned} \tag{8.2}$$

where v_1, v_2, v_3 denote coupling strengths; Coupling matrices $\Delta = (\delta_{ij})_{N\times N}$, $S = (s_{ij})_{N\times N}$, and $R = (r_{ij})_{N\times N}$ are zero-row-sum matrices; Inner coupling matrix $\Gamma = (\gamma_{ij})_{n\times n}$ is an $n \times n$-dimensional matrix; Variable $0 < \omega(t) < \omega$ satisfying $0 < \dot{\omega}(t) < 1$ represents the time-varying delay; $\tilde{\tau} = \max\{\tau, \omega\}$; Design of controller $u_i(t)$ will be discussed soon. Then, denote

$$\Omega_C z_i(t) = [\Omega_{C_1} z_i(t), \Omega_{C_2} z_i(t), \ldots, \Omega_{C_m} z_i(t)]^T,$$

$$\Omega_E z_i(t-\omega(t)) = [\Omega_{E_1} z_i(t-\omega(t)), \Omega_{E_2} z_i(t-\omega(t)), \ldots, \Omega_{E_m} z_i(t-\omega(t))]^T,$$

$$\tilde{\chi}_1(\Omega_C z_i(t)) = [\tilde{\chi}_1^1(\Omega_{C_1} z_i(t)), \tilde{\chi}_2^1(\Omega_{C_2} z_i(t)), \ldots, \tilde{\chi}_m^1(\Omega_{C_m} z_i(t))]^T,$$

$$\tilde{\chi}_2(\Omega_E z_i(t-\omega(t))) = [\tilde{\chi}_1^2(\Omega_{E_1} z_i(t-\omega(t))), \ldots, \tilde{\chi}_m^2(\Omega_{E_m} z_i(t-\omega(t)))]^T,$$

where $\Omega_C = [\Omega_{C_1}, \Omega_{C_2}, \ldots, \Omega_{C_m}]^T$, $\Omega_E = [\Omega_{E_1}, \Omega_{E_2}, \ldots, \Omega_{E_m}]^T$, and $\Omega_{C_k}, \Omega_{E_k} \in \mathbb{R}^n$ with $k = 1, 2, \ldots, m$.

It follows from $e_i(t) = z_i(t) - s(t)$ that

$$\begin{aligned}\dot{e}_i(t) &= \Omega_A e_i(t) + \Omega_B \chi_1(\Omega_C e_i(t)) + \Omega_D \chi_2(\Omega_E e_i(t-\tau(t))) \\ &\quad + v_1\sum_{j=1}^{N}\delta_{ij}\Gamma e_j(t) + v_2\sum_{j=1}^{N} s_{ij}\Gamma e_j(t-\omega(t)) \\ &\quad + v_3\sum_{j=1}^{N} r_{ij}\Gamma \dot{e}_j(t) + u_i(t), \quad i = 1, 2, \ldots, N, \end{aligned} \tag{8.3}$$

where $\chi_1(\Omega_C e_i(t)) = \tilde{\chi}_1(\Omega_C z_i(t)) - \tilde{\chi}_1(\Omega_C z_\pi(t))$, $\chi_2(\Omega_E e_i(t-\tau(t))) = \tilde{\chi}_2(\Omega_E z_i(t-\tau(t))) - \tilde{\chi}_2(\Omega_E z_\pi(t-\tau(t)))$.

To synchronize Lur'e systems in the complex network (8.2) to the target state (8.1), the following impulsive pinning controller is designed as

$$u_i(t) = -\psi_i e_i(t) + \varpi \sum_{k=1}^{+\infty} e_i(t) x(t-t_k), \tag{8.4}$$

where ψ_i represents the pinning strength of the i-th node, ϖ denotes the impulsive effect, $x(t-t_k)$ is the Dirac impulsive function.

Denote $\Psi = \text{diag}\{\psi_1, \psi_2, \ldots, \psi_N\}$ as the pinning gain strength matrix. With $\Omega_A = I_{Nn} \times \Omega_{A_i}$, $\Omega_B = I_{Nn} \times \Omega_{B_i}$, $\Omega_C = I_{Nn} \times \Omega_{C_i}$, $\Omega_D = I_{Nn} \times \Omega_{D_i}$, $\Omega_E = I_{Nn} \times \Omega_{E_i}$, it gives

$$\begin{cases} \dot{e}(t) = \Omega_A e(t) + \Omega_B F_1(\Omega_C e(t)) + \Omega_D F_2(\Omega_E e(t-\tau(t))) \\ \qquad +v_1(\Delta \otimes I_n)e(t) + v_2(S \otimes I_n)e(t-\omega(t)) \\ \qquad +v_3(R \otimes I_n)\dot{e}(t) - (\Psi \otimes I_n)e(t), t \neq t_k, \\ e(t_k^+) = (1+\varpi)e(t_k^-), \quad t = t_k,\ k \in \mathcal{N}^+, \\ e(t) = q(t), \quad t \in [-\tilde{\tau}, 0], \end{cases} \tag{8.5}$$

where $q(t)$ is the initial error value of $e(t)$, $e(t) = [e_1^T(t), e_2^T(t), \ldots, e_N^T(t)]^T$, $F_1(\Omega_C e(t))=[\chi_1^T(\Omega_C e_1(t)), \chi_1^T(\Omega_C e_2(t)), \ldots, \chi_1^T(\Omega_C e_N(t))]^T$, $F_2(\Omega_E e(t-\tau(t))) = [\chi_2^T(\Omega_E e_1(t-\tau(t))), \chi_2^T(\Omega_E e_2(t-\tau(t))), \ldots, \chi_2^T(\Omega_E e_N(t-\tau(t)))]^T$.

Remark 8.1 The synchronization issue discussed in this chapter can also be viewed as a leader-following issue. The target state $z_\pi(t)$ in Eq. (8.1) can be seen as the leader system and the nodes in Lur'e network (8.2) can be regarded as follower systems. Therefore, by constructing the error system (8.5), the synchronization issue has been transformed to the problem of reducing the values of errors $e_i(t)$ to zero.

Remark 8.2 This chapter considers three coupling types for complex networks, i.e., general coupling, time-delay coupling and derivative coupling. Compared to those in [26, 27], the model in this chapter is undoubtedly more comprehensive. For a node in complex networks, the general coupling implies the current state information of other nodes; The time-delay coupling reacts to the information at a certain time ago (caused by transmission delay); The derivative coupling indicates the information of other nodes' sudden changes at the current moment. To explain in detail, if in an n-dimensional space, the displacement of each node/agent is modelled as $z_i^1(t)$, $z_i^2(t)$, ..., $z_i^n(t)$, then the derivative coupling represents their rates of change of velocities, that is, acceleration. Therefore, it is of great significance for judging the relative operating status of other nodes/agents in the complex network.

8.2.2 Related Definitions and Lemmas

Definition 8.1 The synchronization between the i-th Lur'e system $z_i(t)$ and leader system $z_\pi(t)$ is said to be achieved if there exists positive scalars y and φ such that

$$\|z_i(t) - z_\pi(t)\| \leq y\chi(\psi_i(t))e^{-\varphi t}, \ i = 1, 2, \ldots, N$$

holds for any initial values $e_i(t) = \psi_i(t)$, where $\chi(\cdot) : \mathbb{R}^n \to \mathbb{R}^n$ stands for the nonlinear function. It should be mentioned that φ denotes the convergence rate.

Definition 8.2 ([31]) Assume that average impulsive interval is less than T_a, then, for the impulse sequence $\theta = \{t_0, t_1, \ldots\}$, the number of impulsive instants $N_\theta(\bar{t}, t)$ can be estimated by

$$\frac{\bar{t}-t}{T_a} - T_0 \leq N_\theta(\bar{t}, t) \leq \frac{\bar{t}-t}{T_a} + T_0, \quad \forall \bar{t} \geq t \geq 0, \tag{8.6}$$

where constant $T_a, T_0 > 0$.

Lemma 8.1 ([32]) *If there exists a function $\nu(t)$ which satisfies following three conditions: (1) Right-continuous: $\nu_a(t_k) = \nu_a(t_k^+) = \lim_{t\to t_k^+} \nu_a(t)$; (2) Limitation of left-hand exists: $\nu(t_k^-) = \lim_{t\to t_k^-} \nu(t)$ exists; (3) $\nu(t) \in PC([t_0 - \tilde{\tau}, t_0], R)$, then, for two functions $\nu_a(t)$ and $\nu_b(t)$ which have the same property as $\nu(t)$, it can be obtained $\nu_a(t) \leq \nu_b(t)$ for $t \in [0, +\infty)$ if $\nu_a(t) \leq \nu_b(t)$ for $t \in [-\tilde{\tau}, 0]$.*

$$\begin{cases} D^+\nu_a(t) \leq \varrho\nu_a(t) + \varsigma\nu_a(t-\tau(t)), \ t \neq t_k, \\ \nu_a(t_k) \leq w\nu_a(t_k^-), \ k \in \mathcal{N}, \end{cases}$$

$$\begin{cases} D^+\nu_b(t) > \varrho\nu_b(t) + \varsigma\nu_b(t-\tau(t)), \ t \neq t_k, \\ \nu_b(t_k) = w\nu_b(t_k^-), \ k \in \mathcal{N}. \end{cases}$$

Assumption 8.1 The functions $\chi_1(\cdot)$ and $\chi_2(\cdot)$ satisfy the Lipschitz conditions if following two inequalities hold

$$\|\tilde{\chi}_1(\alpha) - \tilde{\chi}_1(\beta)\| \leq \zeta_1\|\alpha - \beta\|,$$
$$\|\tilde{\chi}_2(\alpha) - \tilde{\chi}_2(\beta)\| \leq \zeta_2\|\alpha - \beta\|,$$

where α, β are n-dimensional vectors and ζ_1, ζ_2 are positive constants.

8.3 Main Results

Theorem 8.1 *Suppose that controlled Lur'e dynamical networks (8.5) satisfy Assumption 8.1. For the impulses sequence $\theta = \{t_1, t_2, \ldots\}$, the average impulsive interval is assumed to be less than T_a.*

(Case I). *Consider that impulsive effect $\varpi \in \left(-\infty, -\sqrt{\frac{1}{\lambda_{max}(I_N - \upsilon_3 R)}} - 1\right) \cup \left(\sqrt{\frac{1}{\lambda_{max}(I_N - \upsilon_3 R)}} - 1, +\infty\right)$. If there exists a diagonal matrix $\Psi > 0$ and scalars g, ϱ, $\varsigma > 0$, such that*

$$
\begin{aligned}
&2\Omega_A^s + \Omega_B\Omega_B^T + \zeta_1\Omega_C^T\Omega_C + \Omega_D\Omega_D^T + 2\upsilon_1(\Delta^s \otimes I_n) + \upsilon_2^2 I_{Nn} \\
&\quad - 2\Psi \otimes I_n + g(I_N - \upsilon_3 R) \otimes I_n < 0, &(8.7)\\
&\zeta_2\Omega_E\Omega_E^T - \varrho(I_N - \upsilon_3 R) \otimes I_n < 0, &(8.8)\\
&(S^T \otimes I_n)(S \otimes I_n) - \varsigma(I_N - \upsilon_3 R) \otimes I_n < 0, &(8.9)\\
&\frac{\ln w}{T_a} - g + (\varrho + \varsigma)w^{T_0} < 0, &(8.10)
\end{aligned}
$$

where $w = \lambda_{max}(I_N - \upsilon_3 R)(1+\varpi)^2$, then the controlled Lur'e dynamical networks (8.5) will exponentially achieve stable. Additionally, the exponential convergence rate is specifically estimated as $\frac{\varphi}{2}$, where $\varphi > 0$ is the solution of equation $\varphi - g + \frac{\ln w}{T_a} + \varrho w^{T_0} e^{\tau\varphi} + \varsigma w^{T_0} e^{\omega\varphi} = 0$. That is, the synchronization between leader system (8.1) and the follower systems in Lur'e networks (8.2) is realized via the designed impulsive pinning controller (8.4).

***(Case II)**. Consider that impulsive effect $\varpi \in \Big[-\sqrt{\frac{1}{\lambda_{max}(I_N-\upsilon_3 R)}} - 1, \sqrt{\frac{1}{\lambda_{max}(I_N-\upsilon_3 R)}} - 1\Big]$. If conditions (8.7)–(8.10) in **Case I** are still satisfied and*

$$
\frac{\ln w}{T_a} - g + (\varrho + \varsigma)w^{-T_0} < 0, \tag{8.11}
$$

then the controlled Lur'e dynamical networks (8.5) will exponentially achieve stable. Additionally, the exponential convergence rate is specifically estimated as $\frac{\bar{\varphi}}{2}$, where $\bar{\varphi} > 0$ is the solution of equation $\bar{\varphi} - g + \frac{\ln w}{T_a} + \varrho w^{-T_0} e^{\tau\bar{\varphi}} + \varsigma w^{-T_0} e^{\omega\bar{\varphi}} = 0$. That is, the synchronization between leader system (8.1) and the follower systems in Lur'e networks (8.2) is realized via the designed impulsive pinning controller (8.4).

Proof The Lyapunov function is selected as

$$
V(t) = \frac{1}{2} e(t)^T ((I_N - \upsilon_3 R) \otimes I_n) e(t). \tag{8.12}
$$

For $t \neq t_k$, it can be obtained that

$$
\begin{aligned}
V(t_k^+) &= \frac{1}{2} e^T(t_k^+)((I_N - \upsilon_3 R) \otimes I_n) e(t_k^+) \\
&\le \frac{1}{2}\lambda_{\max}(I_N - \upsilon_3 R)(1+\varpi)^2 e^T(t_k^-) e(t_k^-) \\
&\le \frac{1}{2}(1+\varpi)^2 \frac{\lambda_{\max}(I_N - \upsilon_3 R)}{\lambda_{\min}(I_N - \upsilon_3 R)} e^T(t_k^-)((I_N - \upsilon_3 R) \otimes I_n) e(t_k^-) \\
&\le (1+\varpi)^2 \lambda_{\max}(I_N - \upsilon_3 R) V(t_k^-) = w V(t_k^-).
\end{aligned} \tag{8.13}
$$

For $t \in [t_{k-1}, t_k)$, $k \in \mathcal{N}^+$, with consideration of conditions (8.7)–(8.10) and by calculation of the derivative of (8.12) along the error Lur'e network (8.5), we have

$$
\begin{aligned}
\dot{V}(t) &= e(t)^T[(I_N - v_3 R) \otimes I_n]\dot{e}(t) \\
&\leq \frac{1}{2}e(t)^T(\Omega_A + \Omega_A^T + \Omega_B\Omega_B^T + \zeta_1\Omega_C^T\Omega_C + \Omega_D\Omega_D^T + 2v_1(\Delta \otimes I_n) \\
&\quad + v_2^2 I_{Nn} - 2(\Psi \otimes I_n))e(t) + \frac{1}{2}\zeta_2 e(t-\tau(t))^T\Omega_E\Omega_E^T e(t-\tau(t)) \\
&\quad + \frac{1}{2}e(t-\omega(t))^T(S^T \otimes I_n)(S \otimes I_n)e(t-\omega(t)) \\
&= \frac{1}{2}e(t)^T(\Omega_A + \Omega_A^T + \Omega_B\Omega_B^T + \zeta_1\Omega_C^T\Omega_C + \Omega_D\Omega_D^T + v_1(\Delta \otimes I_n) \\
&\quad + v_1(\Delta^T \otimes I_n) + v_2^2 I_{Nn} + g(I_N - v_3 R) \otimes I_n - 2(\Psi \otimes I_n))e(t) \\
&\quad - \frac{1}{2}g e(t)^T((I_N - v_3 R) \otimes I_n)e(t) \\
&\quad + \frac{1}{2}e(t-\tau(t))^T(\zeta_2\Omega_E\Omega_E^T - \varrho(I_N - v_3 R) \otimes I_n)e(t-\tau(t)) \\
&\quad + \frac{1}{2}e(t-\omega(t))^T((S^T \otimes I_n)(S \otimes I_n) - \varsigma(I_N - v_3 R) \otimes I_n)e(t-\omega(t)) \\
&\quad + \frac{1}{2}\varrho e(t-\tau(t))^T((I_N - v_3 R) \otimes I_n)e(t-\tau(t)) \\
&\quad + \frac{1}{2}\varsigma e(t-\omega(t))^T((I_N - v_3 R) \otimes I_n)e(t-\omega(t)) \\
&\leq -gV(t) + \varrho V(t-\tau(t)) + \varsigma V(t-\omega(t)). \qquad (8.14)
\end{aligned}
$$

With consideration of (8.13), (8.14) and $\epsilon > 0$, the comparison system $v(t)$ is modelled as

$$
\begin{cases}
\dot{v}(t) = -gv(t) + \varrho v(t-\tau(t)) + \varsigma v(t-\omega(t)) + \epsilon, t \neq t_k, \\
v(t_k^+) = wv(t_k^-), \quad k \in \mathcal{N}^+, \\
v(t) = \sum_{i=1}^{N} \|q_i(t)\|^2, \quad t \in [-\tilde{\tau}, 0].
\end{cases}
\qquad (8.15)
$$

From Lemma 8.1, it can be acknowledged that there holds $V(t) \leq v(t)$ when $t < 0$. In term of the extended parameter variation formula, the integral equation of $v(t)$ can be further calculated as

$$
v(t) = \mu(t,0)v(0) + \int_0^t \mu(t,s)(\varrho v(s-\tau(s)) + \varsigma v(s-\omega(s)) + \epsilon)ds, t \geq s \geq 0, \qquad (8.16)
$$

where $\mu(t,s)$ denotes the Cauchy matrix of the system which can be expressed by

$$
\begin{cases}
\dot{v}(t) = -gv(t), \quad t \neq t_k, \\
v(t_k^+) = wv(t_k^-), \quad k \in \mathcal{N}^+.
\end{cases}
$$

In the following discussion, the synchronization conditions and exponential convergence velocities will be derived in regard with two different functions of impulsive effects.

(Case I). For the desynchronizing impulses $w > 1$, i.e., $\varpi(\varpi \neq -1)$ satisfying $\varpi > \sqrt{\frac{1}{\lambda_{\max}(I_N - v_3 R)}} - 1$ or $\varpi < -\sqrt{\frac{1}{\lambda_{\max}(I_N - v_3 R)}} - 1$. With help of average impulsive interval defined in Definition 8.2, we have

$$\begin{aligned}\mu(t,s) &= e^{-g(t-s)} \prod_{s \le t_k \le t} w \le e^{-g(t-s)} w^{T_\theta(t,s)} \\ &\le e^{-g(t-s)} w^{\frac{t-s}{T_a}+T_0} \le w^{T_0} e^{(\frac{\ln w}{T_a}-g)(t-s)}.\end{aligned} \tag{8.17}$$

To simplify the expression, define $\rho = w^{T_0} \sup_{-\tilde{\tau} \le s \le 0}\{\|q(s)\|^2\}$ and $\sigma = g - \frac{\ln w}{T_a}$. By jointly considering (8.16) and (8.17), we can obtain

$$\begin{aligned}v(t) &= \mu(t,0)\cdot v(0) + \int_0^t \mu(t,s)(\varrho v(s-\tau(s)) + \varsigma v(s-\omega(s)) + \epsilon)ds \\ &\le w^{T_0} \sup_{-\tilde{\tau}\le s\le 0} \|q(s)\|^2 e^{(\frac{\ln w}{T_a}-g)t} + \int_0^t e^{(\frac{\ln w}{T_a}-g)(t-s)} w^{T_0} \\ &\quad \times (\varrho v(s-\tau(s) + \varsigma v(s-\omega(s)) + \epsilon)ds \\ &= \rho e^{-\sigma t} + \int_0^t e^{-\sigma(t-s)}(\varrho w^{T_0} v(s-\tau(s)) + \varsigma w^{T_0} v(s-\omega(s)) + w^{T_0}\epsilon)ds.\end{aligned} \tag{8.18}$$

According to (8.2), it can be known that two time-varying delays $\tau(t)$ and $\omega(t)$ exist simultaneously in the complex network. Therefore, a function of variable φ is constructed to help dealing the delay couplings.

$$\varepsilon(\varphi) = \varphi - g + \frac{\ln w}{T_a} + \varrho w^{T_0} e^{\tau\varphi} + \varsigma w^{T_0} e^{\omega\varphi},$$

which is a continuous function for $\varphi \in \mathbb{R}$. By simple calculation, it gives $\varepsilon(0) = -g + \frac{\ln w}{T_a} + (\varrho + \varsigma) w^{T_0} < 0$ and $\dot{\varepsilon}(\varphi) = 1 + \tau\varrho w^{T_0} e^{\tau\varphi} + \omega\varsigma w^{T_0} e^{\omega\varphi} > 0$, which indicates that function $\varepsilon(\varphi)$ is monotonically increasing. As a result, there exists at least a φ^* letting $\varepsilon(\varphi^*) = 0$.

With parameters $w > 1$, $\varphi > 0$, $\varepsilon > 0$ and for any $t \in [-\tilde{\tau}, 0]$, it gives

$$v(t) \le w^{T_0} \sup_{-\tilde{\tau}\le s\le 0} \{\|q(t)\|^2\} < \rho e^{-\varphi t} + \frac{w^{T_0}\epsilon}{\sigma - (\varrho + \varsigma) w^{T_0}}, \tag{8.19}$$

then, for all $t > 0$, we want to prove

$$v(t) < \rho e^{-\varphi t} + \frac{w^{T_0}\epsilon}{\sigma - (\varrho + \varsigma)w^{T_0}}. \tag{8.20}$$

Contradiction method is introduced to illustrate the feasibility of (8.20) when $t > 0$. Firstly, assume that inequality (8.20) is not true for all $t > 0$, then, there at least exists an instant $t > t^*$ satisfying

$$v(t^*) \geq \rho e^{-\varphi t^*} + \frac{w^{T_0}\epsilon}{\sigma - (\varrho + \varsigma)w^{T_0}}, \quad t > t^*. \tag{8.21}$$

In addition, for $t \in [0, t^*)$, inequality (8.20) still holds, namely,

$$v(t) < \rho e^{-\varphi t} + \frac{w^{T_0}\epsilon}{\sigma - (\varrho + \varsigma)w^{T_0}}, \quad t < t^*. \tag{8.22}$$

It follows from (8.18) and (8.22) that

$$\begin{aligned}
v(t^*) &= \mu(t^*, 0) \cdot v(0) + \int_0^{t^*} \mu(t^*, s) \cdot (\varrho v(s - \tau(s)) + \varsigma v(s - \omega(s)) + \epsilon) ds \\
&\leq \rho e^{-\sigma t^*} + \int_0^{t^*} e^{-\sigma(t^* - s)} (\varrho w^{T_0} v(s - \tau(s)) + \varsigma w^{T_0} v(s - \omega(s)) + w^{T_0}\epsilon) ds \\
&\leq \rho e^{-\sigma t^*} + \frac{w^{T_0}\epsilon}{\sigma - (\varrho + \varsigma)w^{T_0}} e^{-\sigma t^*} + e^{-\sigma t^*} \varrho w^{T_0} \rho e^{\tau\varphi} \\
&\quad \int_0^{t^*} e^{(\sigma-\varphi)s} ds + e^{-\sigma t^*} \varsigma w^{T_0} \rho e^{\omega\varphi} \int_0^{t^*} e^{(\sigma-\varphi)s} ds \\
&\quad + e^{-\sigma t^*} \varrho w^{T_0} \frac{w^{T_0}\epsilon}{\sigma - (\varrho + \varsigma)w^{T_0}} \int_0^{t^*} e^{\sigma s} ds + e^{-\sigma t^*} \varsigma w^{T_0} \\
&\quad \times \frac{w^{T_0}\epsilon}{\sigma - (\varrho + \varsigma)w^{T_0}} \int_0^{t^*} e^{\sigma s} ds + w^{T_0}\epsilon e^{-\sigma t^*} \int_0^{t^*} e^{\sigma s} ds \\
&= \rho e^{-\sigma t^*} + \frac{w^{T_0}\epsilon}{\sigma - (\varrho + \varsigma)w^{T_0}} e^{-\sigma t^*} + \frac{\varrho w^{T_0}\rho}{\sigma - \varphi} e^{-\sigma t^*} e^{\tau\varphi} \\
&\quad \times (e^{(\sigma-\varphi)t^*} - 1) + \frac{\varsigma w^{T_0} a}{\sigma - \varphi} e^{-\sigma t^*} e^{\omega\varphi} (e^{(\sigma-\varphi)t^*} - 1) \\
&\quad + \frac{\varrho w^{T_0} w^{T_0}\epsilon}{\sigma(\sigma - (\varrho + \varsigma)w^{T_0})} e^{-\sigma t^*} (e^{\sigma t^*} - 1) \\
&\quad + \frac{\varsigma w^{T_0} w^{T_0}\epsilon}{\sigma(\sigma - (\varrho + \varsigma)w^{T_0})} e^{-\sigma t^*} (e^{\sigma t^*} - 1) + \frac{w^{T_0}\epsilon}{\sigma} e^{-\sigma t^*} (e^{\sigma t^*} - 1) \\
&= \rho e^{-\sigma t^*} + \frac{w^{T_0}\epsilon}{\sigma - (\varrho + \varsigma)w^{T_0}} e^{-\sigma t^*} + \frac{w^{T_0}\epsilon}{\sigma} (1 - e^{-\sigma t^*}) \\
&\quad + \frac{w^{T_0}\rho(q e^{\tau\varphi} + \varsigma e^{\omega\varphi})}{\sigma - \varphi} (e^{-\varphi t^*} - e^{-\sigma t^*})
\end{aligned}$$

$$+\frac{(\varrho+\varsigma)w^{T_0}w^{T_0}\epsilon}{\sigma(\sigma-(\varrho+\varsigma)w^{T_0})}(1-e^{-\sigma t^*}). \tag{8.23}$$

As discussed before that $\varepsilon(\varphi^*)=0$, we have

$$\begin{aligned}
v(t^*) &\le \rho e^{-\sigma t^*}+\frac{w^{T_0}\epsilon}{\sigma-(\varrho+\varsigma)w^{T_0}}e^{-\sigma t^*}\\
&\quad+\frac{\rho(\varrho w^{T_0}e^{\tau\varphi}+\varsigma w^{T_0}e^{\omega\varphi})}{\sigma-\varphi}(e^{-\varphi t^*}-e^{-\sigma t^*})\\
&\quad+\frac{(\varrho+\varsigma)w^{T_0}w^{T_0}\epsilon}{\sigma(\sigma-(\varrho+\varsigma)w^{T_0})}(1-e^{-\sigma t^*})+\frac{w^{T_0}\epsilon}{\sigma}(1-e^{-\sigma t^*})\\
&=\rho e^{-\sigma t^*}+\rho(e^{-\varphi t^*}-e^{-\sigma t^*})+\frac{w^{T_0}\epsilon}{\sigma-(\varrho+\varsigma)w^{T_0}}e^{-\sigma t^*}\\
&\quad+\frac{w^{T_0}\epsilon}{\sigma}(1-e^{-\sigma t^*})+\frac{(\varrho+\varsigma)w^{T_0}w^{T_0}\epsilon}{\sigma(\sigma-(\varrho+\varsigma)w^{T_0})}(1-e^{-\sigma t^*})\\
&=\rho e^{-\varphi t^*}+\frac{w^{T_0}\epsilon}{\sigma-(\varrho+\varsigma)w^{T_0}},
\end{aligned} \tag{8.24}$$

which is contradict to the assumption in (8.21). Therefore, it demonstrates that $v(t)<\rho e^{-\varphi t}+\frac{w^{T_0}\epsilon}{\sigma-(\varrho+\varsigma)w^{T_0}}$ holds for $t\in[-\tilde{\tau},+\infty)$. It yields from $\epsilon\to 0$ that

$$\frac{1}{2}\lambda_{\min}(I_N-v_3R)\|e(t)\|^2\le v(t)<\rho e^{-\varphi t}+\frac{w^{N_0}\epsilon}{\sigma-(\varrho+\varsigma)w^{T_0}}\to\rho e^{-\varphi t}, \tag{8.25}$$

and it can be further calculated as

$$\|e(t)\|\le\sqrt{\frac{2w^{T_0}\sup_{-\tilde{\tau}\le s\le 0}\{\|q(t)\|^2\}}{\lambda_{\min}(I_N-v_3R)}}e^{-\frac{\varphi}{2}t}. \tag{8.26}$$

Above all, it can be acquired that the controlled Lur'e networks (8.5) will finally achieve exponentially stable. In the other word, the exponential synchronization between derivative coupled Lur'e network (8.2) and leader Lur'e network (8.1) is accomplished with utilization of the impulsive pinning controller (8.4) with the case $\varpi\in\left(-\infty,-\sqrt{\frac{1}{\lambda_{\max}(I_N-v_3R)}}-1\right)\cup\left(\sqrt{\frac{1}{\lambda_{\max}(I_N-v_3R)}}-1,+\infty\right)$. Additionally, it should be pointed out that the exponential convergence rate is precisely estimated as $\frac{\varphi}{2}$, where φ is the unique solution of the equation $\varphi-(g-\frac{\ln w}{T_a})+(\varrho e^{\tau\varphi}+\varsigma e^{\omega\varphi})w^{T_0}=0$.

(Case II). For the desynchronizing impulses $0<w\le 1$, i.e., $\varpi(\varpi\ne -1)$ satisfying $-\sqrt{\frac{1}{\lambda_{\max}(I_N-v_3R)}}-1\le\varpi\le\sqrt{\frac{1}{\lambda_{\max}(I_N-v_3R)}}-1$. With help of average impulsive interval defined in Definition 8.2, we have

$$\mu(t,s) = e^{-g(t-s)} \prod_{s \le t_k \le t} w \le e^{-g(t-s)} w^{T_\theta(t,s)}$$
$$\le e^{-g(t-s)} w^{\frac{t-s}{T_a} - T_0} \le w^{-T_0} e^{(\frac{\ln w}{T_a} - g)(t-s)}. \tag{8.27}$$

Denote $\bar{\rho} = w^{-T_0} \sup_{-\bar{\tau} \le s \le 0}\{\|q(s)\|^2\}$. With similar operations in (8.18), substituting (8.27) into (8.16), then

$$v(t) = \mu(t,0) \cdot v(0) + \int_0^t \mu(t,s)(\varrho v(s-\tau(s)) + \varsigma v(s-\omega(s)) + \epsilon) ds$$
$$\le \bar{\rho} e^{-\sigma t} + \int_0^t e^{-\sigma(t-s)}(\varrho w^{-T_0} v(s-\tau(s)) + \varsigma w^{-T_0} v(s-\omega(s)) + w^{T_0}\epsilon) ds. \tag{8.28}$$

Construct another impulsive solution equation in regard to variable $\bar{\varphi}$ by

$$\varepsilon(\bar{\varphi}) = \bar{\varphi} - g + \frac{\ln w}{T_a} + \varrho w^{-T_0} e^{\tau\bar{\varphi}} + \varsigma w^{-T_0} e^{\omega\bar{\varphi}}.$$

In accordance with similar mathematical deductions of Theorem 8.1, for any $t > 0$, it gives

$$v(t) < \bar{\rho} e^{-\bar{\varphi} t} + \frac{w^{-T_0}\epsilon}{\sigma - (\varrho + \varsigma) w^{-T_0}}. \tag{8.29}$$

It yields from $\epsilon \to 0$ that

$$\frac{1}{2}\lambda_{\min}(I_N - v_3 R)\|e(t)\|^2 \le v(t) < \bar{\rho} e^{-\bar{\varphi} t} + \frac{w^{-T_0}\epsilon}{\sigma - (\varrho + \varsigma) w^{-T_0}} \to \bar{\rho} e^{-\bar{\varphi} t}, \tag{8.30}$$

and it can be further derived that

$$\|e(t)\| \le \sqrt{\frac{2 w^{-T_0} \sup_{-\bar{\tau} \le s \le 0}\{\|q(t)\|^2\}}{\lambda_{\min}(I_N - v_3 R)}} e^{-\frac{\bar{\varphi}}{2} t}. \tag{8.31}$$

Above all, it can be acquired that the controlled Lur'e networks (8.5) will finally achieve exponentially stable. In the other word, the exponential synchronization between derivative coupled Lur'e network (8.2) and leader Lur'e network (8.1) is accomplished with utilization of the impulsive pinning controller (8.4) with the case $\varpi \in \left[-\sqrt{\frac{1}{\lambda_{\max}(I_N - v_3 R)}} - 1, \sqrt{\frac{1}{\lambda_{\max}(I_N - v_3 R)}} - 1\right]$. Additionally, it should be pointed out that the exponential convergence rate is precisely estimated as $\frac{\bar{\varphi}}{2}$, where $\bar{\varphi}$ is the unique solution of the equation $\bar{\varphi} - (g - \frac{\ln w}{T_a}) + (\rho e^{\tau\bar{\varphi}} + \varsigma e^{\omega\bar{\varphi}}) w^{-T_0} = 0$. Until now, the proof of Theorem 8.1 is totally finished. □

Remark 8.3 Due to the existence of derivative coupling $c_3\sum_{j=1}^{N} r_{ij}\Gamma\dot{y}_j(t)$, some mathematical techniques are used. By sorting the eigenvalues of matrix as $\lambda_N \geq \cdots \geq \lambda_2 \geq \lambda_1 = 0$, it can be achieved that $\lambda_{\min}(I_N - \upsilon_3 R) = 1$ and $\lambda_{\max}(I_N - \upsilon_3 R) \geq 1$. Through some scaling techniques in (8.13), the relationship between $V(t_k^+)$ and $V(t_k^-)$ is revealed.

Remark 8.4 From Eq. (8.4), it can be acknowledged that control inputs turn out to be a kind of impulsive pinning controllers if taking the effects of continuous control and discrete control into account at the same time, that is, $\psi_i > 0$ and $\varpi \neq 0$. On the contrary, the controller $u_i(t)$ will transfer to simply an impulsive controller if $\psi_i = 0$ or a negative pinning controller if $\varpi = 0$. As demonstrated in Theorem 8.1, when impulses play a positive role in synchronization, the pinning gains of feedback controller will be decreased to save control cost. Oppositely, if impulses are found to be desynchronizing impulses, the control gains ψ_i will enhance to compensate the control effects. Hence, the cost and efficiency will be balanced by the dynamic balance of the pinning strength.

Remark 8.5 Most existed works have excessively restricted the range of the impulsive effects when discussing synchronizing impulses and desynchronizing impulses separately, consequently, leading to the conservativeness of the theorems. In fact, the impulsive effect is meaningless only when taking values 0 or 1. If the impulsive effect is set to $\varpi = 0$, namely, $e(t_k^+) = I_{Nn}e(t_k^-)$, indicating that no change happens at instant t_k; and when the impulsive effect is set to $\varpi = 1$, namely, $e(t_k^+) = 0$, the error turns out to be zero instantly, which is obviously not in line with the reality. However, in [33], the value range of impulsive effects is limited to $(-2, 0)$, thus, the discussion of impulsive effects is not comprehensive. Also, in [34], the impulsive effects only needs to meet $\varpi \neq 0$, which ignoring practical factors. Therefore, this chapter analyzes the synchronizing and desynchronizing impulses by reasonably expanding the value range of impulsive effects, and a less conservative theorem is obtained.

8.4 Numerical Simulation

Three numerical examples will be given in this section to demonstrate the effectiveness of the derived theorem and proposed impulsive pinning controller scheme.

Select the Chua's circuits in [35] as the leader system which is expressed as follows

$$
\begin{cases}
\dot{z}^1(t) = 10(-z^1(t) + z^2(t) + lz^1(t)) \\
\qquad\quad + 0.3247(\|z^1(t) + 1\| - \|z^1(t) - 1\|)) \\
\dot{z}^2(t) = z^1(t) - z^2(t) + z^3(t) \\
\dot{z}^3(t) = -\kappa z^2(t) - 0.1636z^3(t) - \kappa l \sin(0.5z^1(t - \tau(t))),
\end{cases}
\tag{8.32}
$$

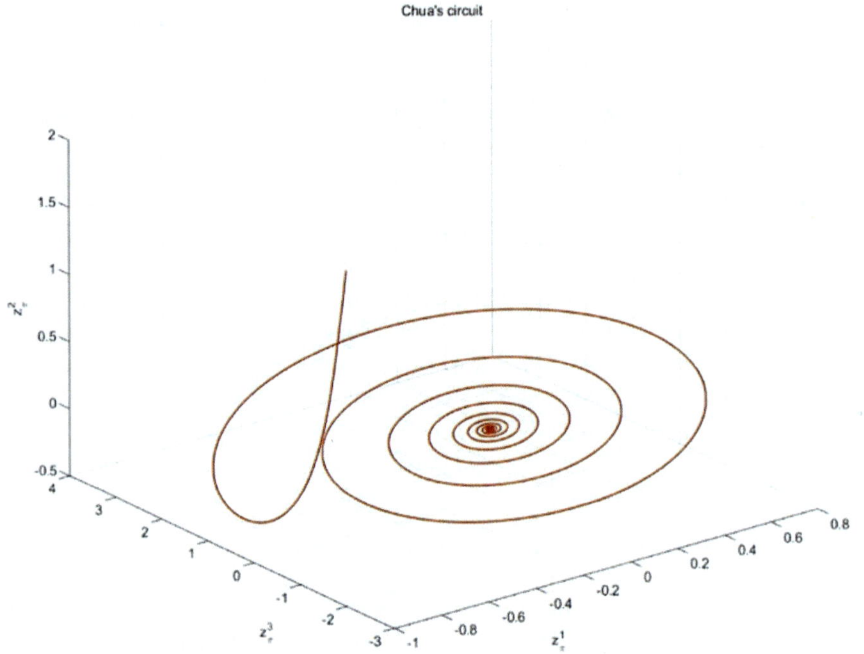

Fig. 8.1 Phase diagram of leader Lur'e network

where $z(t) \in \mathbb{R}^n$ represents the state vector of the Chua's system. Choose the parameters $\kappa = 19.53, l = 0.7831$. Time-varying delay is set as $\tau(t) = \frac{0.01e^t}{1+e^t}$. The trajectory of the state of Lur'e network (8.32) with time-varying delay is displayed in Fig. 8.1. In the following, a Lur'e dynamical network with 6 nodes is presented. Three numerical simulations will be given to show the feasibility of designed control protocol and how the pinning controller behaves when impulsive controller plays different roles in synchronization.

Example 8.1 Let the complex dynamical networks with six Lur'e systems like (8.32). The derivative coupling matrix is given as follows

$$R = \begin{bmatrix} -3 & 1 & 0 & 2 & 0 & 0 \\ 1 & -4 & 1 & 0 & 2 & 0 \\ 0 & 2 & -4 & 1 & 0 & 1 \\ 1 & 1 & 0 & -5 & 2 & 1 \\ 1 & 2 & 2 & 0 & -6 & 1 \\ 0 & 0 & 2 & 1 & 1 & -4 \end{bmatrix}.$$

Choose coupled delay $\omega(t) = 0.1\sin(0.2t)$, coupling strengths $\upsilon_1 = 0.5$, $\upsilon_2 = 0.2$, $\upsilon_3 = 0.5$, and pinning feedback control gain $\psi_i = 0.5$, $(i = 1, \ldots, 6)$. It can be

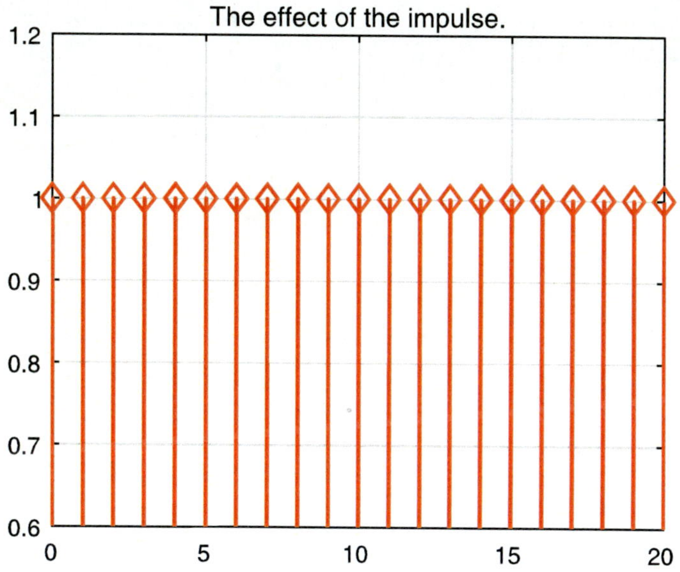

Fig. 8.2 Impulse signals with $T_a = 0.2$

calculated that $\sqrt{\frac{1}{\lambda_{\max}(I_6 - \upsilon_3 R)}} = 0.4739$. Therefore, conditions (8.7)–(8.9) in Theorem 8.1 are obviously satisfied.

Next, the leader-following synchronization issue between follower systems in derivative coupled Lur'e network (8.2) and the leader Lur'e network (8.1) will be verified by applying the designed impulsive pinning controller (8.4). The average impulsive interval is selected as $T_a \leq 0.02$ and the impulsive sequence is denoted by $\theta = \{t_1, t_2, \ldots\}$. The intensity of impulse sequence of impulsive controller is plotted in Fig. 8.2. In this example, impulsive effect is chosen as $\varpi = -2$. By simple calculation, on has $-\sqrt{\frac{1}{\lambda_{\max}(I_6 - \upsilon_3 R)}} - 1 = -1.4739$ and $\sqrt{\frac{1}{\lambda_{\max}(I_6 - \upsilon_3 R)}} - 1 = -0.5261$. Therefore, $\varpi \in (-\infty, -1.4739) \cup (-0.5261, +\infty)$, namely, impulses play a negative role in this example. Moreover, it further gives $w = \lambda_{\max}(I_6 - \upsilon_3 R)(1 + \varpi)^2 > 1$.

By SIMULINK toolbox in MATLAB, the evolution curves of the leader system' state vector $z_\pi(t)$ and the first system in networks $z_1(t)$ are shown in Fig. 8.3a–c. In addition, to show the global information of Lur'e networks, $E^j(t)$ is introduced to be the synchronization error of the j-th state for all Lur'e systems $E^j(t) = \sqrt{\frac{1}{6}\sum_{i=1}^{6}(e_i^j(t))^2}$ with $j = 1, 2, 3$. We give the state curves of synchronization error in Fig. 8.4a–c. In Fig. 8.4, it shows that the synchronization between the follower systems in Lur'e network (8.2) and the leader Lur'e network (8.1) is realized by the designed impulsive controller (8.4).

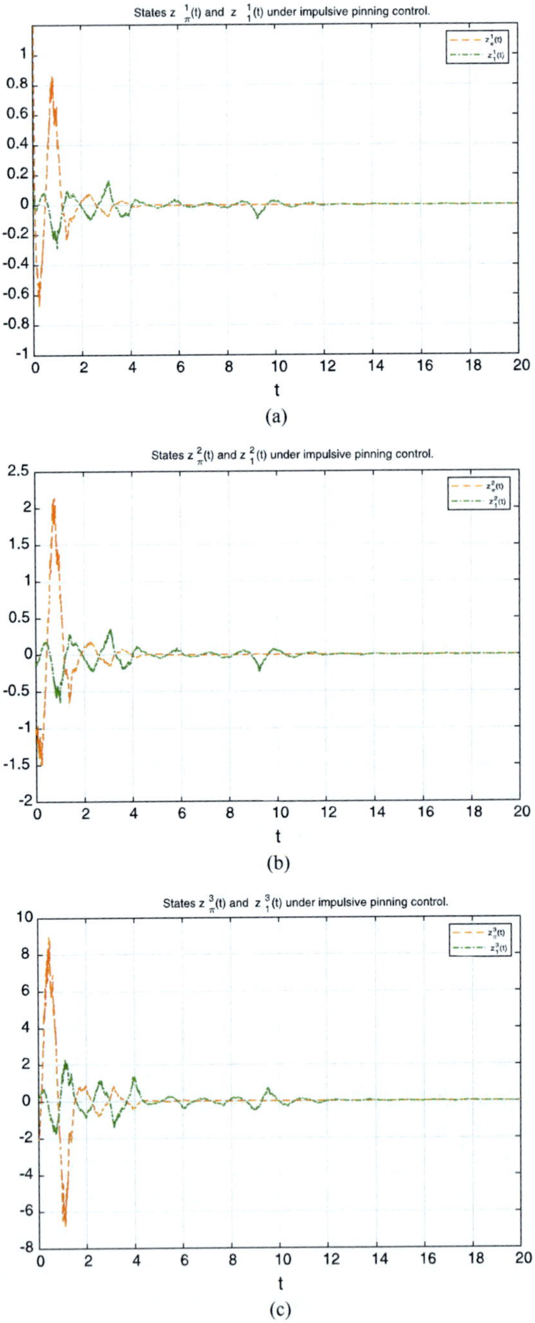

Fig. 8.3 The evolution state curves of $z_{\pi}^{j}(t)$ and $z_{1}^{j}(t)$ with $j = 1, 2, 3$

Fig. 8.4 The evolution curve of the global error $E^j(t)$ with $j = 1, 2, 3$

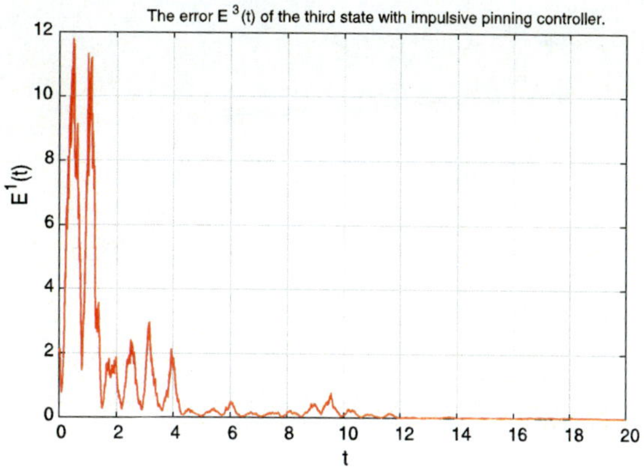

(a)

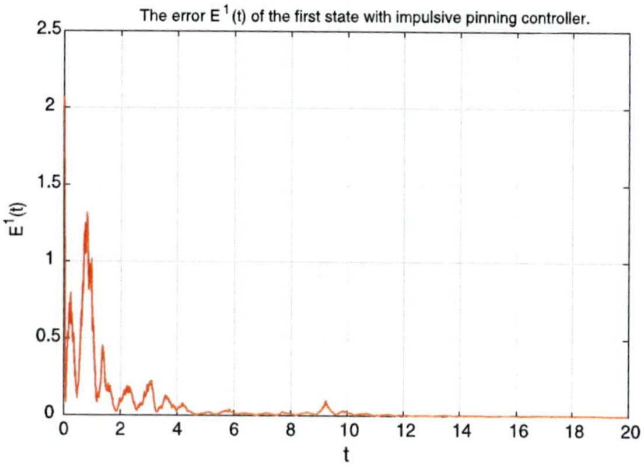

(b)

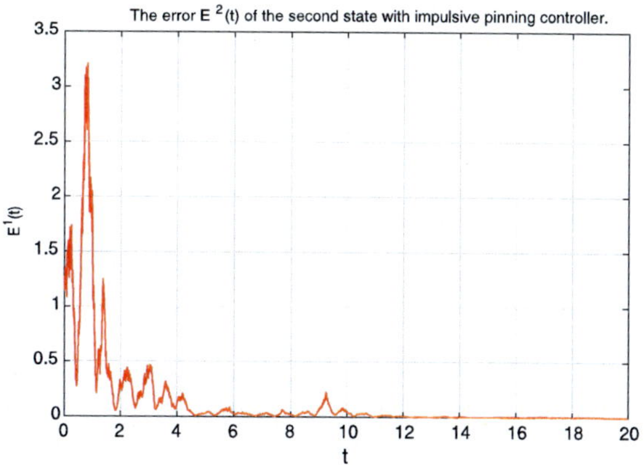

(c)

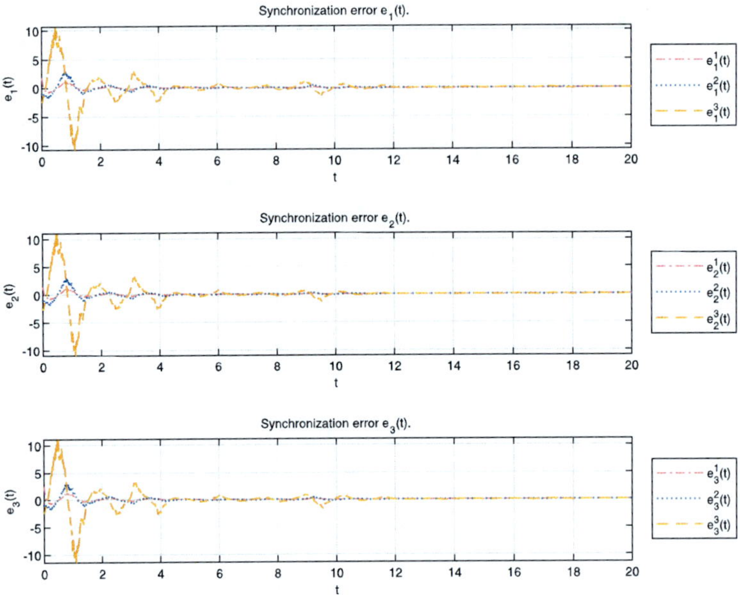

(a)

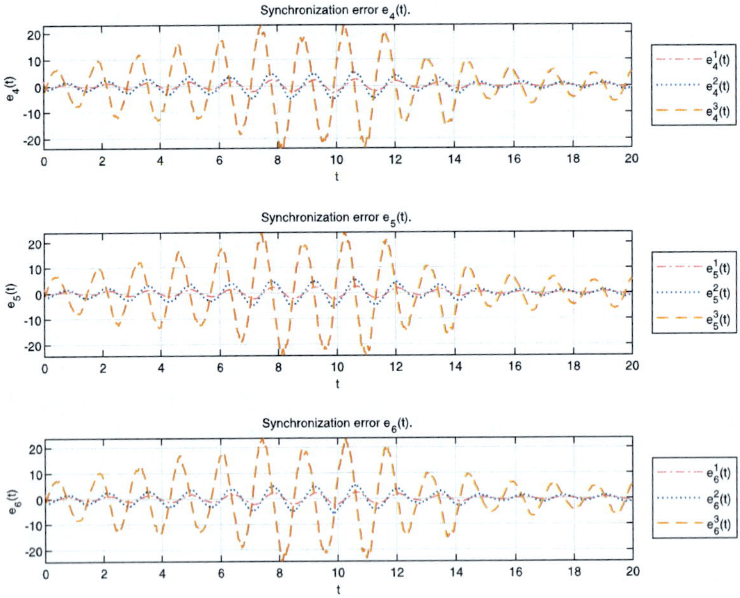

(b)

Fig. 8.5 The evolution curves of error vectors $e_i(t)$ with $\varpi = -0.8$ and $\psi_i = 0.5$, $i = 1, 2, 3, 4, 5, 6$

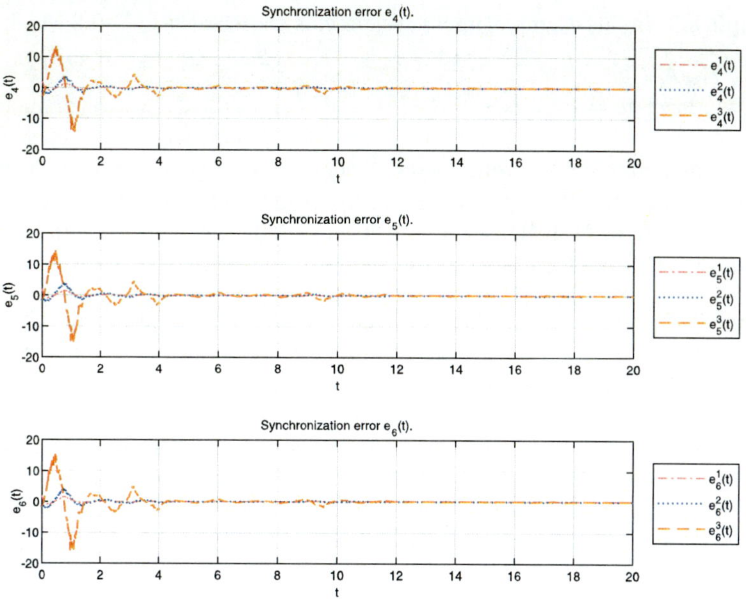

(a)

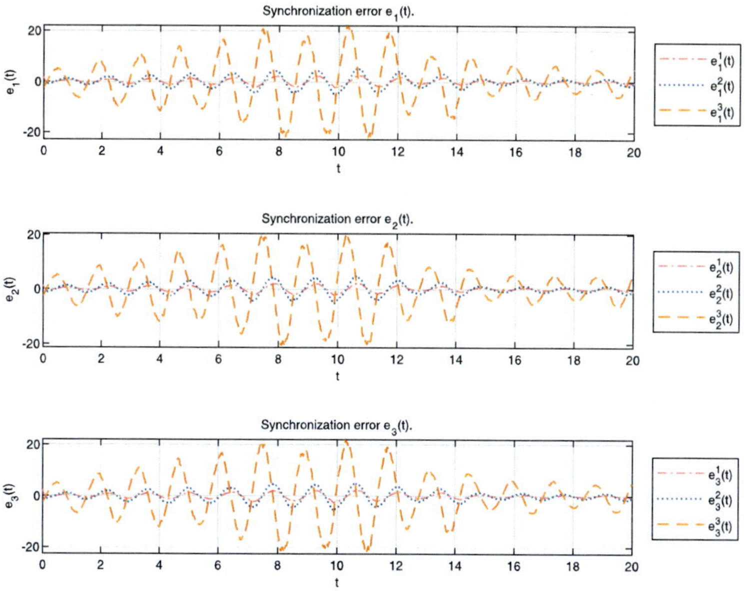

(b)

Fig. 8.6 The evolution curves of error vectors $e_i(t)$ with $\varpi = -0.8$ and $\psi_i = 0, i = 1, 2, 3, 4, 5, 6$

Example 8.2 In this example, impulsive effect is chosen as $\varpi = -0.8$ which satisfies $-\sqrt{\frac{1}{\lambda_{\max}(I_6 - v_3 R)}} - 1 = -1.4739 < \varpi < \sqrt{\frac{1}{\lambda_{\max}(I_6 - v_3 R)}} - 1 = -0.5261$. Accordingly, we have $w = \lambda_{\max}(I_6 - v_3 R)(1 + \varpi)^2 < 1$. With the same parameters in Example 8.1, it is clear that conditions (8.7)–(8.9) are satisfied. Trajectories of error states are plotted in Fig. 8.5 which shows that $e_i(t)$ converge to 0 rapidly. Therefore, it can be obtained that the exponential synchronization between the follower systems in Lur'e network (8.2) and the leader Lur'e network (8.1) is achieved with impulsive effect $\varpi = -0.8$ and feedback control strength $\psi_i = 0.5 (i = 1, \ldots, 6)$.

Example 8.3 In this experiment, the situation where the negative feedback control is selected as $\psi_i = 0$ $(i = 1, \ldots, 6)$ to show the importance of the pinning negative feedback controller part in controller (8.4). At this time, the impulsive gain is $\varpi = -0.8$, which can be viewed as synchronizing impulses. We will study through simulation to see if the synchronization between two Lur'e networks can be realized. In Fig. 8.6, it shows that the states of error vectors cannot approach to zero, namely, the synchronization cannot be achieved without pinning feedback controller. As a conclusion, $\psi_i e_i(t)$ plays an important role in synchronizing the Lur'e dynamical networks.

8.5 Conclusion

A leader-following synchronization issue of Lur'e dynamic networks has been studied in this chapter. To synchronize the follower Lur'e systems in the complex network to the leader Lur'e network, a kind of impulsive pinning control protocols has been established. With help of comparison principle, the definition of average impulsive interval, and Lyapunov stability theorem, synchronization conditions via impulsive controller with different functions of impulsive effects are sufficiently derived. Furthermore, the exponential convergence rates are precisely estimated in regard to synchronizing impulses and desynchronizing impulses. Finally, three numerical simulations are presented to show the feasibility of designed impulsive controller and the necessity of pinning feedback controller.

References

1. Linyuan, L., Zhou, T.: Link prediction in complex networks: A survey. Physica A: Statistical Mechanics and its Applications. **390**, 1150–1170 (2011)
2. Cowan, R., Jonard, N.: Network structure and the diffusion of knowledge. Journal of Economic Dynamics and Control. **28**, 1557–1575 (2004)
3. Boccaletti, S., Latora, V., Moreno, Y., Chavez, M., Hwang, D.U.: Complex networks: Structure and dynamics. Physics Reports. **424**, 175–308 (2006)

4. Soh, H., Lim, S., Zhang, T., Fu, X., Lee, G.K.K., Hung, T.G.G., Di, P., Prakasam, S., Wong, L.: Weighted complex network analysis of travel routes on the Singapore public transportation system. Physica A: Statistical Mechanics and its Applications. **389**, 5852–5863 (2010)
5. Dou, C., Yue, D., Li, X., Xue, Y.: MAS-Based management and control strategies for integrated hybrid energy system. IEEE Transactions on Industrial Informatics. **12**, 1332–1349 (2016)
6. Feng, J., Xie, J., Wang, J., Zhao, Y.: Secure synchronization of stochastic complex networks subject to deception attack with nonidentical nodes and internal disturbance. Information Sciences. **547**, 514–525 (2021)
7. Li, C., Chen, G.: Synchronization in general complex dynamical networks with coupling delays. Physica A: Statistical Mechanics and its Applications. **343**, 263–278 (2004)
8. Liu, X., Chen, T.: Synchronization analysis for nonlinearly-coupled complex networks with an asymmetrical coupling matrix. Physica A: Statistical Mechanics and its Applications. **387**, 4429–4439 (2008)
9. Tan, X., Cao, J., Rutkowski, L.: Distributed dynamic self-triggered control for uncertain complex networks with Markov switching topologies and random time-varying delay. IEEE Transactions on Network Science and Engineering. **7**, 1111–1120 (2020)
10. Long, Y., Park, J.H., Ye, D.: Finite frequency fault detection for a class of nonhomogeneous Markov jump systems with nonlinearities and sensor failures. Nonlinear Dynamics. **96**, 285–299 (2019)
11. Wang, J., Feng, J., Xu, C., Zhao, Y.: Cluster synchronization of nonlinearly-coupled complex networks with nonidentical nodes and asymmetrical coupling matrix. Nonlinear Dynamics. **67**, 1635–1646 (2012)
12. Gao, J., Zhao, Z., Wang, J., Tan, T., Ma, M.: Event-triggered output feedback control for discrete Markov jump systems under deception attack. Journal of the Franklin Institute. **357**, 6435–6452 (2020)
13. Xu, C., Tong, D., Chen, Q., Zhou, W., Xu, Y.: Exponential synchronization of chaotic systems with stochastic noise via periodically intermittent control. International Journal of Robust and Nonlinear Control. **30**, 2611–2624 (2020)
14. Wu, K., Li, B., Du, Y., Du, S.: Synchronization for impulsive hybrid-coupled reaction-diffusion neural networks with time-varying delays. Communications in Nonlinear Science and Numerical Simulation. **82**, 105031 (2020)
15. Yu, C., Qin, J., Gao, H.: Cluster synchronization in directed networks of partial-state coupled linear systems under pinning control. Automatica. **50**, 2341–2349 (2014)
16. Ahmed, M.A.A., Liu, Y., Zhang, W., Alsaadi, F.E.: Exponential synchronization via pinning adaptive control for complex networks of networks with time delays. Neurocomputing. **225**, 198–204 (2017)
17. Cheng, L., Yang, Y., Li, L., Sui, X.: Finite-time hybrid projective synchronization of the drive-response complex networks with distributed-delay via adaptive intermittent control. Physica A: Statistical Mechanics and its Applications. **500**, 273–286 (2018)
18. Zhou, P., Cai, S., Shen, J., Liu, Z.: Adaptive exponential cluster synchronization in colored community networks via a periodically intermittent pinning control. Nonlinear Dynamics. **92**, 905–921 (2018)
19. Zhang, B., Dou, C., Yue, D., Zhang, Z., Zhang, T.: A cyber-physical cooperative hierarchical control strategy for islanded microgrid facing with random communication failure. IEEE Systems Journal. **14**, 2849–2860 (2020)
20. Liu, D., Ye, D.: Pinning-observer-based secure synchronization control for complex dynamical networks subject to DoS attacks. IEEE Transactions on Circuits and Systems I: Regular Papers. **67**, 5394–5404 (2020)
21. Peng, D., Li, X.: Leader-following synchronization of complex dynamic networks via event-triggered impulsive control. Neurocomputing. **412**, 1–10 (2020)
22. Li, H., Hu, C., Jiang, Y., Wang, Z., Teng, Z.: Pinning adaptive and impulsive synchronization of fractional-order complex dynamical networks. Chaos, Solitons and Fractals. **92**, 142–149 (2016)

23. Li, Z., Fang, J., Huang, T., Miao, Q.: Synchronization of stochastic discrete-time complex networks with partial mixed impulsive effects. Journal of the Franklin Institute. **354**, 4196–4214 (2017)
24. Zhou, Y., Zeng, Z.: Event-triggered impulsive control on quasi-synchronization of memristive neural networks with time-varying delays. Neural Networks. **110**, 55–65 (2019)
25. He, W., Qian, F., Cao, J.: Pinning-controlled synchronization of delayed neural networks with distributed-delay coupling via impulsive control. Neural Networks. **85**, 1–9 (2017)
26. Wang, Y., Lu, J., Liang, J., Cao, J., Perc, M.: Pinning synchronization of nonlinear coupled Lur'e networks under hybrid impulses. IEEE Transactions on Circuits and Systems II: Express Briefs. **66**, 432–436 (2019)
27. Wu, Y., Li, Y., Li, W.: Synchronization of random coupling delayed complex networks with random and adaptive coupling strength. Nonlinear Dynamics. **96**, 2393–2412 (2019)
28. Chen, H., Shi, P., Lim, C.C.: Synchronization control for neutral stochastic delay Markov networks via single pinning impulsive strategy. IEEE Transactions on Systems, Man, and Cybernetics: Systems. **50**, 5406–5419 (2020)
29. Cai, S., Zhou, P., Liu, Z.: Synchronization analysis of hybrid-coupled delayed dynamical networks with impulsive effects: A unified synchronization criterion. Journal of the Franklin Institute. **352**, 2065–2089 (2015)
30. Yu, N., Zhu, W.: Event-triggered impulsive chaotic synchronization of fractional-order differential systems. Applied Mathematics and Computation. **388**, 125554 (2021)
31. Lu, J., Ho, D.W.C., Cao, J.: A unified synchronization criterion for impulsive dynamical networks. Automatica. **46**, 1215–1221 (2010)
32. Yang, Z., Xu, D.: Stability analysis and design of impulsive control systems with time delay. IEEE Transactions on Automatic Control. **52**, 1448–1454 (2007)
33. Wu, Z., Liu, D., Ye, Q.: Pinning impulsive synchronization of complex-variable dynamical network. Communications in Nonlinear Science and Numerical Simulation. **20**, 273–280 (2015)
34. Wang, X., Liu, X., She, K., Zhong, S.: Pinning impulsive synchronization of complex dynamical networks with various time-varying delay sizes. Nonlinear Analysis: Hybrid Systems. **26**, 307–318 (2017)
35. Tang, Z., Park, J.H., Lee, T.H.: Distributed adaptive pinning control for cluster synchronization of nonlinearly coupled Lur'e networks. Communications in Nonlinear Science and Numerical Simulation. **39**, 7–20 (2016)